ENGINEERING CERAMICS: FABRICATION SCIENCE & TECHNOLOGY

Edited by
D. P. THOMPSON
*Materials Division, Department of Mechanical, Materials &
Manufacturing Engineering, The University of
Newcastle upon Tyne, NE1 7RU*

**British Ceramic Proceedings
No.50**

THE INSTITUTE OF MATERIALS

Book 563
Published in 1993 by
The Institute of Materials
1 Carlton House Terrace
London SW1Y 5DB

ISBN 0-0901716-40-5

The papers contained in this fiftieth volume
of the British Ceramic Proceedings were presented
at a meeting of the Basic Science Section
of The Institute of Ceramics
held at Manchester Business School,
18–20 December, 1991.

Typeset, printed and bound
in Great Britain by
Sherwin Rivers Ltd.
Cobridge, Stoke-on-Trent

Contents

Plasma Synthesis and Processing of Ultra-Fine Ceramic Powders

G. P. DRANSFIELD

Tioxide Specialties Ltd., Haverton Hill Road, Billingham, Cleveland, TS23 1PS, UK

ABSTRACT

Vapour phase synthesis of ceramic powders has attracted much interest in the literature. Most of this has concerned the preparation of very small amounts in the laboratory. Large scale production is, however possible. Tioxide use DC plasma synthesis to make ≈ 50,000 tpa of pigmentary titanium dioxide of crystal size ~ 0·2 μm. This technology has now been applied to the preparation of ultra-fine oxide and non-oxide ceramic powders. Furthermore, it has been shown that pigment coating technology is an excellent method of distributing additives homogeneously. Coating has led to discernible benefits in terms of sintering and mechanical properties of the fired ceramic.

1. INTRODUCTION

This paper will describe ways of making ultra-fine oxide and non-oxide powders by DC plasma synthesis and subsequent powder treatment. The properties of the resultant powders, sintering characteristics and the attributes of the resultant ceramics will also be reported.

2. EXPERIMENTAL

2.1 Preparation of Oxide Powders

Tioxide's plasma technology has been used to manufacture a number of oxide powders. Zirconia powder with a crystal size of ca. 70 nm has been made by plasma oxidation of zirconium tetrachloride at temperatures above 1200°C.

$$ZrCl_4 + O_2 = ZrO_2 + 2 Cl_2$$

Using patented methods, the zirconia can then be coated with stabilizers such as yttria, ceria and magnesia [1, 2]. The production route outlined in Figure 1 is essentially similar to that for "chloride route" titania pigment.

The resultant powder is very fine, having a mean size of ~ 0·07 μm. Figure 2 illustrates the size distribution and morphology of the base zirconia powder.

2.2 Preparation on Non-oxide Powders

Gas phase routes are being used by Tioxide for the production of nitrides, for instance titanium nitride.

There are three main routes to titanium nitride, similar to those for any non-oxide material.

1. Direct nitridation of the element. In the case of titanium this is in the form of either powder or sponge.

$$Ti + 1/2 N_2 \text{ (or NH}_3) = TiN \tag{1}$$

2. Carbothermal reduction of the nitride.

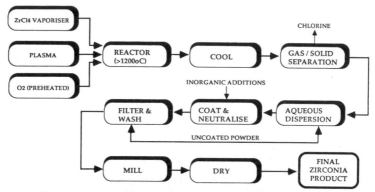

Figure 1. Process flow chart for yttria coated zirconia powder.

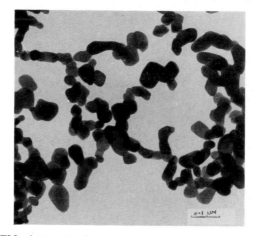

Figure 2. TEM micrograph of zirconia powder, produced by plasma oxidation.

$$TiO_2 + 2\,C + 1/2\,N_2 \; (or\; NH_3) = TiN + 2\,CO \qquad (2)$$

3. Vapour phase reaction of the chloride with ammonia.

$$TiCl_4 + 2\,NH_3 = TiN + 4\,HCl + 1/2\,N_2 + H_2 \qquad (3)$$

The powder produced by the third method is substantially finer than TiN produced by more conventional methods (1 and 2). This is illustrated in Figure 3.

The size of the powder can be controlled using a patented reactor recirculation method to be in the range 10 to 200 nm, as desired [3]. Powder varying in colour from black to brown can be obtained in this manner. This gives rise to interesting pigmentary properties.

A potential problem with such ultra-fine powders is high oxygen content. Nitrides are thermodynamically unstable with respect to oxides and hence each individual particle gives a surface coating of oxide on exposure to air. It has been demonstrated that this oxygen is present at the surface since there is a linear correlation between the specific surface area and oxygen content (Figure 4) [4].

It is possible for any given application to establish an optimum in the trade-off between particle size and oxygen content and powder can be tailor-made to this specification.

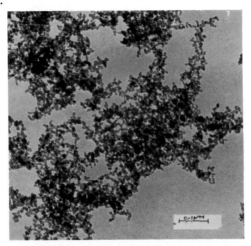

Figure 3. TiN powder prepared by DC plasma vapour phase synthesis.

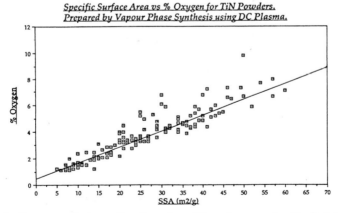

Figure 4. Oxygen content as a function of specific surface area for titanium nitride.

3. POWDER CHARACTERISTICS

3.1 Oxide Powders

Yttria stabilized forms of zirconia can be sintered at temperatures as low as 1350°C to near theoretical density [5]. The resulting ceramics have a fine grain size, as illustrated in Figure 5.

Because of the fine, uniform microstructure, sintered ceramic pieces have excellent mechanical strength, up to 1·5 GPa, when measured by the biaxial disc testing technique of Sivill [6].

One of the major drawbacks of yttria stabilized zirconia is its susceptibility to degradation by water vapour in the range 100–300°C. This phenomenon, known as *ageing*, is a major factor limiting widespread use and can be referred to as the "Achilles heel" of yttria stabilized zirconia. Tioxide derived material is known to have remarkably superior resistance to ageing [7]. This was demonstrated when sintered discs of this material and materials prepared by a chemical precipitation route were aged at 180°C in an autoclave for 200 hours according to the method of Nakajima *et al.* [8]. The results are shown below in graphical form in Figure 6.

It is believed that the use of a coating to introduce the yttria stabilizer is responsible for the superior ageing properties of the Tioxide material. Other

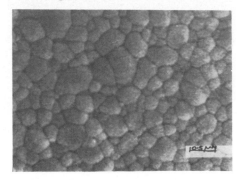

Figure 5. Microstructure of Tioxide Y-TZP sintered at 1450°C.

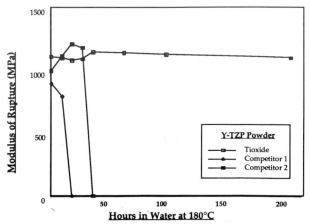

Figure 6. Ageing properties of Tioxide coated Y-TZP compared with two equivalent co-precipitated powders.

promising features of yttria coated zirconia include: high green strength (> 20 MPa), high sintered fracture toughness and good ionic conductivity.

It has also been shown that coating zirconia with ceria leads to exceptionally high strengths (up to 1·25 GPa) [9]. Stabilization is possible at ceria levels significantly lower than the minimum value of 8 mol% reported by Tsukuma *et al.* [10]. Other rare earth oxides can also be added as a coating to control the microstructure. In particular, the addition of neodymia has led to a fine, uniformly-grained microstructure and enables further reduction in the required level of ceria, as demonstrated by Figures 7(a) and 7(b).

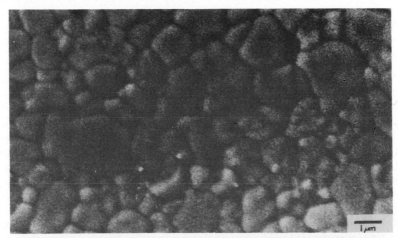

Figure 7(a). Microstructure of a sintered 7 mol% CeO₂ coated ZrO₂ powder.

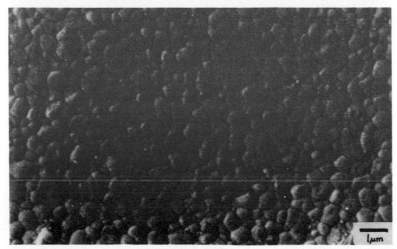

Figure 7(b). Microstructure of a sintered 4·5 mol% CeO₂ + 1 mol% Nd₂O₃ co-coated ZrO₂ powder.

A composition containing 5·7 mol% CeO_2 + 1 mol% Nd_2O_3 had excellent mechanical properties: a hardness of 10 GPa, fracture toughness of 15 MPa m$^{1/2}$ and strength of 1·25 GPa.

Commercial quantities of zirconia powder are now being made by gas phase synthesis using Tioxide's proprietary plasma technology. Tioxide are now supplying a range of yttria stabilized powders containing between 2 and 8 mol% (a further advantage of coating is that it allows total flexibility with regard to additive level). The ultra-pure, additive-free zirconia powder may also be used as a precursor for electroceramic materials. A range of other zirconia products are being planned such as ceria stabilized zirconia, zirconia based composites and powders tailored to specific needs, *e.g.* a powder which can be readily injection moulded. These issues are being addressed through Tioxide's participation in British Government and European Community sponsored collaborative projects covering oxides and non-oxides.

3.2 Non-oxide Powders

The ultra-fine titanium nitride made by the gas phase route shows a number of advantages over powders obtained by other routes. Titanium nitride is an electrically conductive powder and can be used as an electrically conducting additive to silicon nitride or other ceramic powders. This can render the sintered ceramic piece electro-discharge machinable, thereby considerably reducing the costs of ceramic component manufacture. Using an ultra-fine titanium nitride powder, conductivity can be achieved at 30% rather than 50% by weight loadings required for the coarser, competitor powders. This is illustrated in Figure 8.

This enhanced conductivity is presumably due to a more efficient dispersion of the fine particles through the insulating matrix. Plasma derived titanium

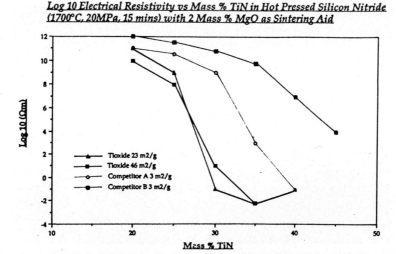

Figure 8. Electrical resistivity vs. mass% TiN in hot pressed silicon nitride.

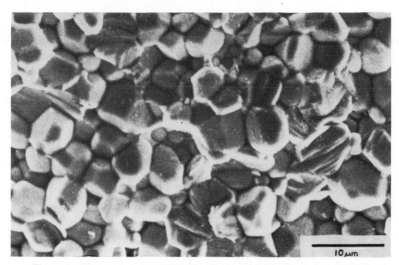

Figure 9. Microstructure of Tioxide titanium nitride sintered at 1600°C.

nitride powders show excellent sintering properties. They can be pressureless sintered without sintering aids at temperatures as low as 1400°C. The density achieved is dependent on the surface area of the powder and reaches an optimum around 25 m²/g. The sintered product has a very fine grain structure, as illustrated in Figure 9.

The sintered strength approaches 500 MPa, hardness is ~ 15 GPa and the fracture toughness is around 4 MPa m$^{1/2}$. Sintered pieces are electrically conducting and the conductivity is, in fact, higher than that of metallic titanium [11]. Commercial quantitities of the powder are now being made with surface areas ranging from 5 to 70 m²/g.

It has been shown that Tioxide's plasma technology can be used to synthesize other nitrides; especially those of silicon and aluminium. The resultant powders are white in colour and have a particle size of ~ 50 nm. The silicon nitride has an amorphous structure. Preliminary measurements have been encouraging. The powders show superior sintering characteristics to powders made by more conventional means in much the same way that titanium nitride does. High strengths (exceeding 800 MPa) have already been achieved on sintering the silicon nitride powder [12] and commercially significant quantities of this powder will be produced during 1992. Plasma derived aluminium nitride powder is still at the development stage. However, aluminium nitride is important as a substrate material and the availability of an improved starting powder represents a significant development in this field.

4. CONCLUSIONS

The viability of plasma synthesis to make ultra-fine ceramic powders of high purity and narrow size distribution on an industrial scale has been successfully demonstrated. Furthermore, other aspects of pigment manufacturing technology, such as the application of coatings; can also be used to make powders which display unique advantages over those made by more conventional means. Tioxide are now in the early stages of commercialisation and hope ultimately to supply powders over the whole spectrum of advanced ceramic application, using their expertise in gas phase synthesis, applying nanometre scale coatings, dispersing powders and controlling aggregation.

ACKNOWLEDGMENTS

The author wishes to thank Tioxide Group Ltd. for their kind permission to publish this article. The author acknowledges the contributions of his colleagues at Tioxide in the preparation of this article.

REFERENCES

1. EGERTON, T. A., LAWSON, J. & FROST, P. W., British Patent Application 2,181,723A, (1986).
2. EGERTON, T. A., FOTHERGILL, K. A. & DRANSFIELD, G. P., British Patent Application 2,204,030A, (1988).
3. EGERTON, T. A., JONES, A. G. & BLACKBURN, S. R., U.K. Patent Application 2,217,699A, (1988).
4. BLACKBURN, S. R., EGERTON, T. A. & JONES, A. G., *Brit. Ceram. Proc.,* **47,** *Fine Ceramic Powders,* Eds. R. Freer and J. L. Woodhead, 87, (1991).
5. DRANSFIELD, G. P., FOTHERGILL, K. A. & EGERTON, T. A., in *EuroCeramics,* **1,** Eds. G. de With, R. Terpstra and R. Metselaar, publ. Elsevier, 275, (1989).
6. SIVILL, A. D., Ph.D. Thesis, Department of Mechanical Engineering, Nottingham University, (1974).
7. DRANSFIELD, G. P., FOTHERGILL, K. A. & EGERTON, T. A., in *Proc. 7th CIMTEC World Ceramics Congress,* Ed. P. Vincenzini, publ. Elsevier (Amsterdam), (1991).
8. NAKAJIMA, K., KOBAYASHI, K. & MURATA, Y., in *Advances in Ceramics, Volume 12, Science and Technology of Zirconia II,* Eds. N. Claussen, M. Ruehle and A. H. Heuer, publ. The American Ceramic Society, Columbus, Ohio, 399, (1984).
9. McCOLGAN, P., DRANSFIELD, G. P. & EGERTON, T. A., The effect of rare earth oxide additions on the mechanical properties of Ce-TZP, to be published in *Proc. 2nd European Ceramic Society Conference,* Augsburg, September, (1991).
10. TSUKUMA, K. & SHIMADA, M., *Am. Ceram. Soc. Bull.,* **65,** 1386, (1986).
11. DRANSFIELD, G. P. & JONES, A. G., *Ibid 9.*
12. HARMSWORTH, P. D., JONES, A. G., EGERTON, T. A. & BLACKBURN, S. R., Gas phase production of a silicon nitride using a DC plasma and powder characteristics, to be published in *Proc. 4th International Symposium on Ceramic Materials and Components for Engines,* Goteborg, Sweden, June, (1991).

The Rheology and Defects of Fibre Loaded Ceramic Extrudates

S. BLACKBURN

*Interdisciplinary Research Centre in Materials for High Performance Applications
and the School of Chemical Engineering
The University of Birmingham, Edgbaston, Birmingham, B15 2TT, UK*

ABSTRACT

*The possibilities of manufacturing fibre loaded ceramic composites by extrusion is
discussed in terms of their rheology and defect forms. A model paste composed of
zirconia and carbon fibre is used to demonstrate the application of a physically based
paste flow model. The analysis shows that the model can be used for short fibres
composite paste systems. With increasing fibre length there is an increased yield stress
and data interpretation becomes less reliable as a result of inhomogeneity introduced
through mixing. Flow instabilities are discussed and related to the rheological data and
to the die design.*

1.INTRODUCTION

It is generally accepted that the strength and fracture toughness of monolithic
ceramics can be improved by the inclusion of relatively short chopped fibres of
the appropriate properties. Such ceramic composites can be formed by many
routes and among these is extrusion. Extrusion is a process whereby the
particulate material is mixed with a suitable liquid to form a compound which
is capable of deformation and flow under pressure. In the case of a composite
material the particulate matrix and the reinforcing material are combined with
the liquid. In extrusion the liquid phase can be a complex combination of
binder, lubricant and solvent. While extrusion is a widely used technique for
the formation of monolithic ceramics and fibre reinforced plastic there is
relatively little data available on the extrusion of fibre loaded ceramic pastes in
the literature. In this paper the rheological behaviour and defect forms of a
model short fibre composite paste comprising ceria stabilised zirconia (Ce-
TZP) and chopped carbon fibres are examined.

1.1 Paste Rheology

A clear understanding of the rheological behaviour of particulate ceramic
pastes has become increasingly important as the complexity and value of the
components being manufactured by extrusion has increased. The development
of physically based extrusion models commenced in 1937, when Berghaus [1]
considered the plastic flow of clay in the manufacture of pipes basing his
theories on the plastic flow of metals. The approach was furthered by Griffith
[2] and more recently advanced by Benbow *et al.* [3]. The essentials of this
approach to modelling paste flow are outlined below:

It has been demonstrated that for low viscosity fluids and particulate mixtures
the pressure drop for paste passing through a simple die can be divided into

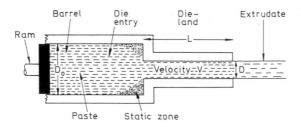

Figure 1. Ram extruder.

two parts, that pressure drop attributable to convergent flow in the die entry (P_1) and that due to sliding along the die land (P_2). P_1 being derived from the theory of metals deformation and P_2 assumes plug flow prevails in the die land. Thus

$$P = P_1 + P_2 = 2[\ln(D_0/D)](\sigma_0) + 4(L/D)(\tau_0 + \beta V) \qquad (1)$$

where σ_0 is the yield stress, τ_0 is the die wall shear stress and β is a factor indicating the change in wall shear stress with extrudate velocity (V). D_0 is the barrel diameter, D is the die diameter and L is the die length, Figure 1. For most paste systems the fluids used are of much higher viscosity and this simple model has been modified by the introduction of an αV term to indicate the variation in yield stress with velocity evident in such systems. Thus, Equation (1) then becomes

$$P = 2[\ln(D_0/D)](\sigma_0 + \alpha V) + 4(L/D)(\tau_0 + \beta V) \qquad (2)$$

where there are now four independent paste parameters. In traditional ceramics extrusion clay suspensions have been widely used as the viscous fluid but with modern technical ceramics clean grain boundaries are often required and polymer based fluids are preferred. In some cases such polymers are highly shear thinning and the pressure drop can often be more closely modelled by the introduction of power indices m and n. Equation (1) becomes [4]

$$P = 2[\ln(D_0/D)](\sigma_0 + \alpha_1 V^m) + 4(L/D)(\tau_0 + \beta_1 V^n) \qquad (3)$$

Table 1. Paste rheology for complex geometries

1. Conical Die Entry Region
$$P = 2[\ln(D_0/D)](\sigma_0 + \alpha V) + 2\cot\gamma\,\tau_0 \ln(D_0/D) + \beta V\cot\gamma \qquad [5]$$

2. Multi Hole Dies
$$P = 2[\ln(D_0/\sqrt{x\{D\}})](\sigma_0 + \alpha V) + 4(L/D)(\tau_0 + \beta V) \qquad [4]$$

3. Dies of Irregular Shape
$$P = [\ln(A_0/A)](\sigma_0 + \alpha V) + 4(LM/A)(\tau_0 + \beta V) \qquad [4]$$

Where γ is the die cone semi angle, x is the number of holes, A is the die open area, A_0 is the barrel area, M is the die perimeter length.

Equations (2) and (3) are for simple circular geometries with square entries. A range of equations has been developed for more complex geometries, these are summarised in Table 1 and are based on Equation (2).

2. EXPERIMENTAL PROCEDURE

2.1 Raw Materials

A zirconia powder stabilised with 15·6 wt% cerium oxide* with a median particle size of 0·62 μm† was used as the particulate material. This material has a narrow size distribution, and comprises discrete dense grains of high sphericity but low roundness. The carbon fibres‡ were selected as a model fibre material for cost and ready availability. The continuous fibres were milled and screened into three different length fractions, −45 (L), 45–63 (B) and 63–90 (A) μm. Due to the inefficiencies of sieving fibres their length distribution was determined by an optical measurement and counting method [6]. During milling the surface roughness of the fibres increased and the diameter of the fibre reduced from typically 7·3 to 6·8 μm. The length distribution of the fibre was determined after extrusion by dissolving the extrudate in water and repeating the optical sizing. The liquid used in all the pastes was a 6 wt% methyl cellulose solution.

2.2 Paste Preparation

Monolithic pastes with moisture contents from 13·9 to 17·3 wt% were prepared by high shear kneading§ for 30 minutes. The composite pastes were prepared by the same route with 20 vol% fibre loading in the solid phase and a moisture range of 13·0 to 16·2 wt%.

2.3 Paste Evaluation

The pastes were extruded from a ram extruder with barrel diameter 25·4 mm and stroke of 150 mm. For rheological analysis dies of 3·175 mm diameter and length 3·175, 25·4, 50·8 mm were employed in the velocity range 0·001 m/s to 0·053 m/s. For defect evaluation a range of dies of different diameters and lengths were used over a similar velocity range. In this paper the four parameter model has been applied throughout.

The extrudates were dried in air at room temperature and impregnated with a low viscosity epoxy resin¶ to facilitate microscopic examination of polished sections.

3. RESULTS AND DISCUSSION

3.1 Powder and Fibre Size Distributions

The zirconia powder size is shown in Figure 2 and the fibre length distributions before mixing and after extrusion are shown in Figure 3. Following extrusion the aspect ratio of the fibres appears to have increased primarily due to

---

* Ce-PSZ-2 μm, Unitec Ceramics, † Measured by Sedigraph 5100, ‡ Courtaulds Fibres.

§ Werner & Pfleiderer, ¶ Logitec 301.

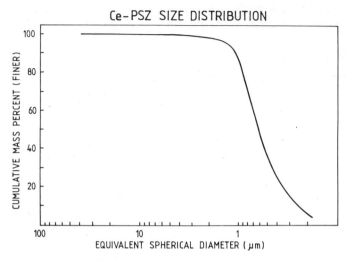

Figure 2. Zirconia size distribution.

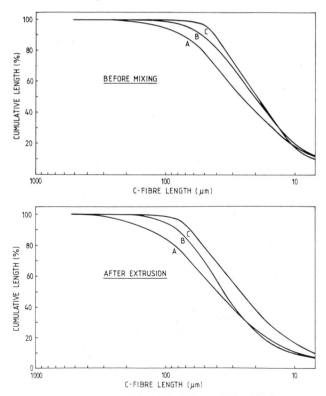

Figure 3. Carbon fibre length distribution.

difficulties in differentiating the short carbon fibres from the zirconia matrix. However, it is clear that the differential between the fibre length distributions remained.

3.2 Rheological Results

The variation in σ_o, τ_o, α and β with moisture content is shown in Figure 4. All four parameters rapidly increase as the moisture content is reduced below 14·5 wt%. This represents the point below which the liquid content is less than the pore volume of the powder and interparticulate friction remains too high for paste flow. The sharpness of the flow transition is indicative of powders of fine size and narrow size distribution. As the liquid volume is reduced the wall shear stress increases relative to the yield value, σ_o/τ_o is reduced from 18 to 8 over the extrudable range investigated. This is attributed to break down of the lubricating liquid layer between the body of the paste moving in plug flow and the die wall. Pastes extruded at the lower limits of moisture content show

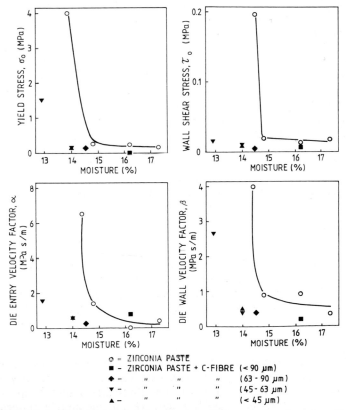

o = ZIRCONIA PASTE
■ - ZIRCONIA PASTE + C-FIBRE (< 90 μm)
♦ -   "   "   "   (63 - 90 μm)
▼ -   "   "   "   (45 - 63 μm)
▲ -   "   "   "   (< 45 μm)

Figure 4. Rheological parameters for zirconia monolithic and zirconia carbon fibre composite pastes.

discolouration from die wear supporting the hypotheses that the lubricating layer is much reduced causing the particulate material to come in direct contact with the die wall.

The rheological data for the composite pastes is shown in Figure 4. The composite pastes consistently gave lower extrusion pressure drops than monolithic pastes of the same weight percent moisture content and this is generally reflected in the four paste parameters. From the currently available data it can be shown that the σ_o/τ_o ratio increases from 10 to 30 as the mean aspect ratio increases from three to five, again with constant moisture. This occurs because increasing the fibre aspect ratio increases σ_o, while τ_o remains constant under uniform liquid conditions. This is consistent with greater resistance to plastic/extensional flow in the die entry region yet constant die wall fibre interaction, both being due to the fibre alignment process. The structure of the dried composite material is shown in Figures 5 and 6. It is clear that there is strong but not perfect alignment of the fibres parallel to extrudate flow. The alignment of the fibres comes about by rotation in the die entry under the influence of shear and extensional flow. Observation of extrudates at the die exit after cutting and the examination of extrudates from dies of different L/D ratios suggests that plug flow occurs in the die land and that no further contribution to fibre alignment occurs in that region. The reduction in the pressure drop for a given moisture content can be attributed to the change in solids density brought about by the addition of carbon fibre to the zirconia and changes in the packing behaviour of the two components [7]. Further, there is possibly a contribution from the carbon fines acting as a lubricant.

The theoretical models described in the introduction appear to hold for the fibre size distributions tested at the 20 vol% loading over the range of moistures examined at low L/D ratios. Some difficulties were experienced with high L/D ratios in pastes containing the 63–90 mm fraction. This may represent flow instabilities reflecting mixing problems due to fibre-fibre interactions.

3.3 Extrudate Evaluation

3.3.1 Shrinkage

During drying there was very little shrinkage in either the monolithic paste or the composite paste. The linear shrinkages are shown in Table 2 for both materials. Such low shrinkages are consistent with the paste being extruded just above the critical moisture content where all porosity is just filled with the binder or liquid phase.

3.3.2 Defects

In composite pastes where the matrix shows a significant shrinkage flaws develop perpendicular to the direction of extrudate flow during drying [8]. These defects come about because of the fibre-fibre interactions and the differential shrinkage between the fibre network and the matrix material. The small differences in shrinkage observed in the zirconia carbon composite

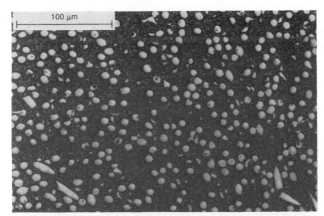

Figure 5. Dried zirconia carbon composite (section, perpendicular to paste flow).

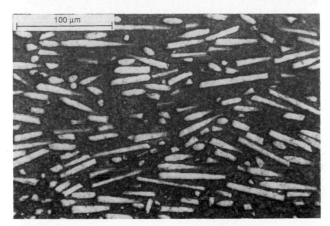

Figure 6. Dried zirconia carbon composite (section, parallel to paste flow).

prevent the development of stresses between the aligned fibres and the matrix during drying and thus this defect form was not observed.

It has been shown [8] that fibre aspect ratio and volume loading affect the homogeneity of paste and that in extreme cases fibre balls can form during mixing. The polished sections of the green extrudate show that in these short chopped fibre highly viscous systems the fibre distribution is reasonably uniform, no fibre balling was observed.

In the monolithic extrudates the nature of the defects changed with moisture content. At high moisture levels the extrudates were smooth and relative to the other pastes defect free. Surface defects first increased as moisture was reduced but then lessened once more to be replaced by cracks perpendicular to the direction of extrusion in the driest materials. This trend was reflected in the green strengths of these materials measured in three point bend where the values were 3·14 and 8·79 MPa for the low and high moisture containing

Table 2. Length shrinkage data

Powder (Fibre length)	Zr			Zr + C fibre		
				C	B	A
Moisture (wt%)	14·4	14·8	16·2	14·0	14·0	14·5
Length shrinkage	0·7	1·1	1·8	0·8	0·8	0·9

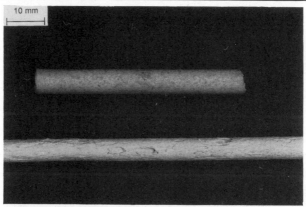

Figure 7. Surface defects (top: typical surface 14·0 wt% moisture, bottom: surface with high moisture content 16·2 wt%), Fibre B.

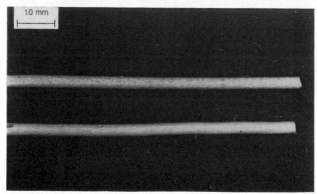

Figure 8. Surface defects (top: L/D = 1, bottom: L/D = 16, both at 14 wt% moisture), Fibre B.

materials respectively, yet after sintering the densities of the materials were all over 99% theoretical density for a Ce-TZP.

In the composite material a different range of defects was observed. At high moisture levels (+16 wt%) the surface of the extrudate showed tearing or "slip-stick" appearing as though internal shearing was occuring possibly indicating departure from pure plug flow (Figure 7) associated with very low yield values. At all other moisture levels examined the extrudates showed clean

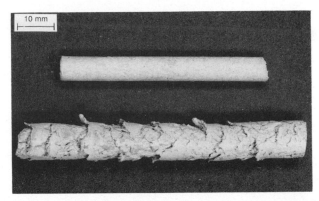

Figure 9. Surface defects (top: typical surface at 14·0 wt% moisture L = 25·4 mm, D = 6 mm, bottom: surface at 14·0 wt% moisture L = 0·5 mm, D = 12 mm), Fibre B.

surfaces when formed through dies with high D_o/D and L/D ratios. There was some slight textural change with die length, Figure 8, where carbon fibre appears to have been drawn to the surface in short die extrusion. It has been shown [9] that in terms of die design short, non-conical entry dies with low D_o/D ratios produce extrudates of inferior surface quality. In these composite pastes this trend was clearly observed, Figure 9. In all the fibre loaded pastes air entrapment increased compared to that of the monolithic pastes, particularly the pastes containing the longest fibres.

4. CONCLUSIONS

The use of physically based rheological models to characterise the paste flow for both monolithic zirconia pastes and carbon fibre-zirconia composite materials has been discussed. It can be concluded that the addition of short fibres reduces the pressure drop required to extrude paste with a fixed weight percentage moisture content. Increasing fibre aspect ratio increases the yield stress relative to the wall shear stress at similar fixed moisture level due to greater force being required to align the longer fibres during convergent flow thus decreasing the plasticity of the paste.

The introduction of the carbon fibres to the monolithic zirconia paste causes only slight changes to extrudate surface instability. Of particular note is the surface "stick-slip" tearing observed at high moisture contents.

By extruding materials just below the minimum possible moisture content the shrinkage during drying is very low and therefore when combined with fibres the internal stresses between the fibre network and matrix were small and internal fractures were absent.

ACKNOWLEDGMENTS

Thanks are due to Mrs. H. Böhm for her invaluable help in the preparation of this paper and to the SERC for their financial support.

REFERENCES

1. BERGHAUS, M. J., *Forsch. geb. Ingenieurwes.*, 23, (1937).
2. GRIFFITHS, R. M., *Can. J. Chem. Eng.*, **44**, 108, (1966).
3. BENBOW, J. J., OXLEY, E. W. & BRIDGWATER, J., *Chem. Eng. Sci.*, **42**, 2151, (1987).
4. BENBOW, J. J., JAZAYERI, S. H. & BRIDGWATER, J., *Powder Technology*, **65**, 393, (1991).
5. BENBOW, J. J., *Chem. Eng. Sci.*, **26**, 1467, (1971).
6. BLACKBURN, S. & LAWSON, T. A., *J. Amer. Ceram. Soc.*, in press.
7. STEDMAN, S. J., EVANS, J. R. G. & WOODTHORPE, J., *J. Mat. Sci.*, **25**, 1833, (1990).
8. BLACKBURN, S., Proc. 2nd European Ceramic Society Conference, Augsburg, (1991), in press.
9. BENBOW, J. J., BLACKBURN, S., LAWSON, T. A., OXLEY, E. W. & BRIDGWATER, J., *Proc. Brit. Ceram. Soc., Special Ceramics,* **9**, (1990), in press.

Production of X-Phase and Mullite/X-Phase Ceramic Powders

C. C. ANYA and A. HENDRY

University of Strathclyde, Department of Metallurgy & Engineering Materials, Glasgow, G1 1XN

ABSTRACT

X-phase is described as having a point composition and its production as a pure material is reported to be difficult. Until now, the stable form was always obtained by hot pressing component oxide/nitride mixtures in the SiO_2-Al_2O_3-AlN-Si_3N_4 system at high temperatures (1700–1780°C).

In this work it is shown that by carbothermally reducing and nitriding a single natural precursor (kaolinite), at the relatively low temperature of 1500°C, 100% X-phase and bi-phase X-phase/mullite powder products can be produced. The advantage of this method lies in its cost effectiveness and ability to produce powders in large quantities.

The paper will describe the method employed in producing these materials from commercial kaolinites of different grades and discusses the influence of processing parameters on the yield and quality of the product. The results of sintering experiments on these materials will also be discussed.

1. INTRODUCTION

In most previous reports [1–4], the formation of X-phase sialon was observed while hot pressing components in the SiO_2-Al_2O_3-AlN-Si_3N_4 system at temperatures between 1700 and 1780°C. The phase has been described as having a range of compositions (Figure 1 [5]), while in some other instances [3, 4, 6] it is said to have a point composition. However, a common feature with the various descriptions is that it is found at equivalent percent of aluminium (eq% Al) values greater than that of kaolinite.

The phase was produced [6] as an intermediate during the production of β'-sialon (z = 3) by carbothermal reduction and nitriding of kaolinite at 1450°C. Kokmeijer *et al.* [7] using the same carbothermal route, and about the same amount of carbon as used by Higgins and Hendry [6], did not observe any X-phase. However, whereas the alumina/silica weight ratios of the kaolinites used by Kokmeijer *et al.* [7] were 1:1·07, 1:1·34, 1:1·36, that used by Higgins and Hendry [6] had a ratio of 1:1·25. Recently, it was reported [8] that the stable form of the phase can be obtained by the carbothermal route at 1500°C, and that its domain of existence as a monocrystalline product is very narrow and includes 42·9 to 49·6 eq% Al.

Apart from the green composition, previous reports [6, 7, 9] have shown that carbothermal synthesis of ceramics depends also on other factors such as nitrogen flow rate, heating time, thoroughness of mixing of green components and granule size/surface area of reactants.

The narrow domain of existence of X as a stable phase [8] and all the variables listed above pose difficulty in trying to obtain the phase as a monocrystalline product. The relative cheapness of the carbothermal route

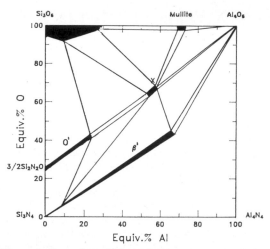

Figure 1. Part of the SiO$_2$-Al$_2$O$_3$-Si$_3$N$_4$-AlN system [5].

coupled with the fact that a single precursor (mixed with carbon) is used, which therefore reduces the number of variables, could make the route a more viable one for producing X-phase in a large quantity. This work is part of a wider programme to produce X-phase and more importantly, X-phase/mullite composite ceramic components. A detailed study of how the above listed factors affect the production of X-phase and mullite/X-phase composite powders through carbothermal synthesis is presented here. Work is continuing on the sintering of the powders so produced, but a short discussion on sintering of the X-phase/mullite composite powder is also presented.

2. EXPERIMENTAL

Details of the experimental procedure were given in an earlier report [8] by the authors. Additional to that, three types of kaolinite and kyanite powders were mixed with the required [8] amount of carbon (Table 1) for the formation of 100% crystalline product of X-phase. Of the three kaolinites, one was the natural kaolinite of alumina/silica weight ratio of 1:1·25 and the

Table 1. Composition of different green mixtures

Batch No.	C* wt%	Al$_2$O$_3$† wt%	SiO$_2$† wt%	Kaolinite† wt%	Kyanite wt%
1.	6	4	—	96	—
2.	6	—	—	94	—
3.	6	—	1·8	98·2	—
4.	6	—	—	—	94

*: The carbon content is relative to the entire green mixture.
†: The alumina, silica and kaolinite (where mixed with alumina or silica) contents are relative to the sum of their partial concentrations considered as 100%.

other two were made up by enrichening the kaolinite with 4 wt% alumina and 1·8 wt% precipitated silica (as a proportion of kaolinite and alumina or silica only) respectively, giving alumina/silica ratios greater and less than 1:1·25. The new ratios were calculated and found to fall out with those used by Kokmeijer *et al.* [7]. The batch with kyanite (alumina/silica ratio of 1:0·6) was made in view of the current accepted position of X-phase in the SiO_2-Al_2O_3-AlN-Si_3N_4 system.

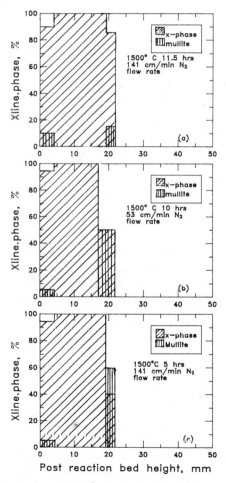

Figure 2. Histogram showing the formation of crystalline phases at various heights of post-reaction bed processed at 1500°C, (a) with N_2 flow rate (NFR) of 141 cm/min. for 11·5 hours, (b) with NFR of 53 cm/min. for 11·5 hours, (c) with NFR of 141 cm/min. for 5 hours.

Pressureless sintering was carried out in a vacuum furnace, which was evacuated prior to heating the sample and nitrogen was allowed to flow continuously throughout the sintering experiment. Exhaust gases at both carbothermal and sintering stages were monitored with an infra-red gas analyser for the presence of CO.

3. RESULTS AND DISCUSSION

3.1 Effect of mixing time of green components, N_2 flow rate and heating time on the proportion of phases

The product profile of post-reaction bed heights of all the batches revealed that the products are in zones, with the middle zone always being the richest in X-phase. Figure 2 shows histograms of the product profile of charges that gave a bi-phase product of X-phase and mullite in varying proportions. Some alumina is observed at the base of the bed but the alumina presence in this case is quite different from that when the green composition is not such as to give 100% crystalline product of X-phase. In the latter, α-Al_2O_3 is present from the bottom to the top zone, as shown in the next section.

Previous reports [6, 7, 9] on carbothermal reduction and nitriding of kaolinite for producing β'-sialon or Si_3N_4 indicate that a copious flow of nitrogen through the bed is critically important for the overall kinetics of the process. The histogram in Figure 2(a) shows the distribution of mullite and X-phase within a post-reaction bed heated for 11·5 hours at 1500°C with a nitrogen flow rate of 141 cm/minute. Figure 2(b) is the same green composition heated under the same conditions as that of Figure 2(a), but with a nitrogen flow rate of 53 cm/minute. Though Figure 2(b) shows more X-phase at the bottom zone, on average, it shows a smaller proportion of X-phase in comparison with that shown in Figure 2(a). The higher proportion of X-phase at the bottom zone of the run with low flow rate is due to an increase in the contact time between nitrogen and oxides. However, the flow not being forceful enough, there was a limited permeability of gases to the top zone. This limited permeability of gases led to the reduced amount of X-phase in this zone. Nevertheless, there seems to be a limit to which the nitrogen flow rate can be increased with a concomitant increase in the amount of X-phase produced. This is because the product profile of a similar green composition processed with the same parameters, but with a nitrogen flow rate of 88 cm/minute was the same as that of 141 cm/minute shown in Figure 2(a).

Figure 2(c) shows the histogram for the same charge heated with a flow rate of 141 cm/minute for 5 hours at 1500°C. Comparing Figure 2(c) with Figure 2(a), it can be seen that on increasing the time of heating from 5 hours to 11·5 hours, the proportion of X-phase increased by about 40%.

It was observed that a mixing time (of the green components) of 6 to 8 hours is sufficient to achieve the maximum yield of X-phase in a carbothermally produced composite powder of mullite/X-phase. Mixing times less than 6 hours were found to lead to the appearance of β'-sialon in the bottom zone.

3.2 Effect of type of kaolinite

The middle zones of batches 1 to 4 were used to generate the results in Figures 3 and 4, which are the low angle section of X-ray diffraction (XRD) patterns of products of the batches in Table 1 heated for 5 hours at 1500°C with a nitrogen flow rate of 141 cm/minute. The crystalline products of batch number 1, Figure 3(a), were X-phase (about 90% — visual estimation), mullite and a trace of β'-sialon. Those of batch number 2, Figure 2(b), were about 96% X-phase and 4% mullite, with no trace of β'-sialon. Batch number 3, Figure 3(c), gave 100% X-phase as a crystalline product, while the product of batch number 4, Figure 4, was heterogeneous, comprising 45% X-phase, 30% β'-sialon, 15% α-Al_2O_3 and 10% mullite.

The results in Figure 3 demonstrate that the type of kaolinite affects the production of X-phase and therefore may partly explain why Kokmeijer *et al.* [7] did not observe any X-phase. Contrary to the general view [3, 4, 6], it appears that X-phase does not have a point composition but exists with a range of composition, as earlier suggested by Jack [5]. The 4% mullite in batch number 2, Figure 3(b), transformed to X-phase on heating for a longer time. In contrast, on heating batch number 1 for a longer time, α-Al_2O_3 was formed with the amount of mullite reduced. Thus, kaolinites of even high alumina content would progressively show reduced amounts of X-phase. Figure 4 also demonstrates that a monocrystalline domain of X-phase does not occur at 60 eq% Al, the value shown by kyanite. Recent work on pressureless sintering and hot isostatic pressing of mixtures of SiO_2, Al_2O_3, Si_3N_4 and AlN at 1700 to 1775°C by Bergman *et al.* [10] also came to the same conclusion that X-phase

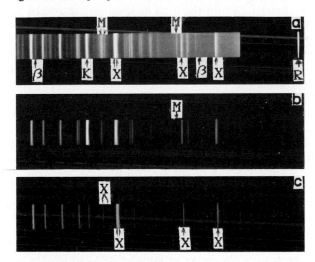

Figure 3. Low angle section of XRD patterns of (a) Batch No. 1, (b) Batch No. 2 and (c) Batch No. 3. Only a few of the principal reflections are labelled in this and subsequent similar figures (X = X-phase, M = mullite, α = α-Al_2O_3, β = β'-sialon, R = reference line and K = KCl). Each batch was heated for 5 hours at 1500°C with NFR of 141 cm/min.

does not have a point composition, but rather could exist as a single crystalline phase between about 41 to 57·8 eq% Al. At 60 eq% Al, the eq% Al value of kyanite, the product was also heterogeneous [10].

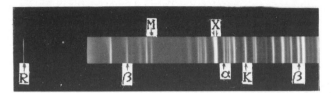

Figure 4. XRD patterns of carbothermally processed raw kyanite (batch No. 4 in Table 1) heated under the same conditions as those in Figure 3.

3.3 Effect of surface area of reactants and heating rate

The efficient carbothermal synthesis of a material requires the surface area of reactants to be high, if a good contact between them is to be achieved. It was suggested [11] that a higher heating rate increases the specific surface area of kaolinite heated up to 1145°C. Therefore, a higher heating rate could be favourable to the overall kinetics of X-phase formation since such a rate would lead to a better contact of reactants. Hence, more X-phase should be expected with a higher heating rate for a given heating period.

Two heating rates, 53 and 14°C/minute were used to heat two charges having all other process parameters the same. After 5 hours of heating, the amounts of X- and mullite phases formed were the same for both heating rates. However, the charge with the high heating rate showed a higher peak of CO, Figure 5. This situation is quite understandable since a higher heating rate implies an increased thermal flux which would be expected to cover a wider reaction front, thus facilitating greater evolution of CO for a given time. The figures inside the graph are the temperatures, in degrees centigrade at which these peaks were observed.

The specific area of carbon is more effective than that of the oxide in accelerating transformations in carbothermal synthesis; in fact, with carbon of a high surface area, the oxide surface area has no effect at all [9]. The carbon used here is of a very high surface area — 1500 m²/g, which explains why a higher heating rate, though producing oxides of high surface areas did not have any effect on the kinetics of the overall process.

Worthy of note in Figure 5 are firstly, that CO evolution starts from around 100°C and stops after about 2 hours, the time being shorter for higher heating rates. This CO evolves at low temperatures by reaction with free moisture in the kaolinite; the powder and crucible, prior to heating were dried at 160°C for 24 hours, so the possibility of adsorbed water reacting with carbon is ruled out. Secondly, except for the peaks at about 500°C which are associated [6] with reaction of volatiles in the carbon with bonded water from kaolinite, the other peaks did not characterise the stages of the transformation. The peaks usually appeared about a minute after the charge was advanced towards the

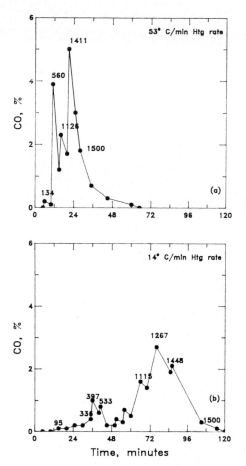

Figure 5. CO evolution versus heating time showing various peaks at specified temperatures (in °C) for heating rates of: (a) 53°C/min., (b) 14°C/min.

hot zone, each time by an equal distance. Thus, the peaks appear when a new reaction front is created within the bed, and the highest peak appears when the time of creating the new front coincides with that of attaining a temperature with substantially high thermal flux. Subsequent high temperatures, inclusive of that of mullite/X-phase transformation do not lead to higher peaks since by then, virtually all the charge has become a reaction front.

3.4 Sintering of carbothermally produced X-phase/mullite powder

Two-phase composite powders of about 50% mullite/50% X-phase produced according to the process described above were sintered at 1590°C for 35 minutes. Further CO evolution (maximum of about 0·8%) started at about 1250°C. Figure 6 shows the XRD patterns of the powder in unsintered

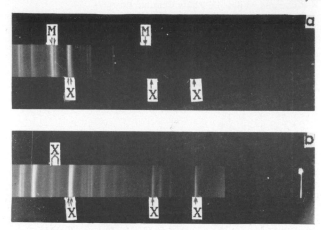

Figure 6. XRD patterns of 50% mullite/50% X-phase in, (a) unsintered and (b) sintered forms.

(Figure 6(a)) and sintered (Figure 6(b)) forms. It can be seen that all the mullite transformed to X-phase. Hence, the failure of the reaction at carbothermal stage to proceed to 100% crystalline X-phase product within the entire bed-height, even though the green composition was so designed, most probably is due to a chemical equilibrium effect. At higher temperature of sintering, the overall composition moves to higher X-phase compositions.

Further evolution of CO at the sintering temperature is an indication that all of the carbon was not burnt at the carbothermal stage. The higher temperature at the sintering stage may have led to a new equilibrium point, thus facilitating the oxidation of this residual carbon and hence further transformation of mullite to X-phase. The relative ease of the transformation could further be explained if the congruent melting of X-phase is considered. It has been suggested [4] that X-phase melts congruently and that its melting point is below 1750°C [3, 10]. As the heating temperature is increased, the sample gets nearer to the melting temperature of X-phase, which in this case could be below 1750°C due to impurities in the kaolinite and the presence of mullite. Hence, the smallest degree of undercooling would lead to the formation of a very substantial amount of X-phase.

It follows therefore that a sintered product of X-phase and mullite cannot be made from a composite powder of the two produced simultaneously by a carbothermal process of green compositions designed to give 100% X-phase crystalline product. A possible way to make a sintered mullite/X-phase composite from such a powder may be to use green compositions (at the carbothermal stage) that have a lesser amount of carbon, corresponding to a slightly higher equivalent percent of oxygen (eq% O) than that present in X-phase.

The sintered product was also very porous, Figure 7. This may be a combined effect of escaping gas bubbles within the temperature range of mullite → X-phase transformation and the transformation itself. A two-step

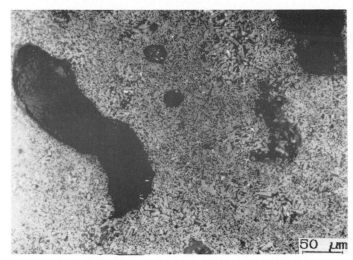

Figure 7. Optical micrographs of carbothermally produced 50% mullite/50% X-phase samples sintered at 1590°C for 35 minutes. After sintering, all the mullite transformed to X-phase. Note the large pores.

sintering, with the first step at a lower temperature and longer than the second step may be of help in reducing the porosity.

4. CONCLUSIONS

In the process of forming X-phase by carbothermal reduction and nitriding of kaolinite in a packed bed, products are formed in zones, the middle zone being the richest in X-phase. Providing the correct green composition is used, a longer heating time leads to more X-phase being formed, but 100% of the bed height is never converted entirely to X-phase at 1500°C.

A mixing time (of green components) of 6 to 8 hours and a nitrogen flow rate of 90 to 100 cm/minute are sufficient to yield the maximum concentration of X-phase within a pre-reaction bed height of about 35 mm (with a bed diameter of 32 mm) at 1500°C.

Higher heating rates did not have any effect on the kinetics of mullite/X-phase transformation, though they always led to higher peaks of CO. These peaks were observed to appear mainly as new reaction fronts were created within the bed.

A complete transformation of mullite to X-phase takes place when a composite of the two, simultaneously produced by a carbothermal process of a green mixture designed to give 100% X-phase, is sintered at temperatures higher than those of the carbothermal process.

ACKNOWLEDGMENT

Financial support under BRITE-EURAM programme (contract number, Breu-0064-EDB) is acknowledged.

REFERENCES

1. ZANGVIL, A., *J. Mat. Sci. Lett.*, **13**, 1370, (1978).
2. ZANGVIL, A., GAUCKLER, L. J. & RUHLE, M., *idem*, **15**, 788, (1980).
3. NAIK, I. K., GAUCKLER, L. J. & TIEN, T. Y., *J. Amer. Ceram. Soc.*, **61**, 332, (1978).
4. THOMPSON, D. P. & KORGUL, P., in 2nd NATO ASI Conf. Proc., *Progress in Nitrogen Ceramics,* Ed. F. L. Riley, publ. M. Nijhoff, 375, (1983).
5. JACK, K. H., *J. Mat. Sci.*, **11**, 1135, (1976).
6. HIGGINS, I. & HENDRY, A., *Brit. Ceram. Trans. & J.*, **85**, 161, (1986).
7. KOKMEIJER, E., SCHOLTE, C., BLOMER, F. & METSELAAR, R., *J. Mat. Sci.*, **25**, 1261. (1990).
8. ANYA, C. C. & HENDRY, A., *J. Eur. Ceram. Soc.*, **10**, 65, (1992).
9. DURHAM, S. J. P., SHANKER, K. & DREW, R. A. L., *J. Amer. Ceram. Soc.*, **74**, 31, (1991).
10. BERGMAN, B., EKSTRÖM, T. & MISCKI, A., *J. Eur. Ceram. Soc.*, **8**, 141, (1991).
11. BLESCO, A., BARBA, A., NEGRE, F. & ESCARDINO, A., *Brit. Ceram. Trans. & J.*, **89**, 28, (1990).

Modelling of Self-Propagating High-Temperature Synthesis Reactions

C. R. BOWEN and B. DERBY

Department of Materials, University of Oxford, Parks Road, OX1 3PH, UK

ABSTRACT

The modelling of self-propagating high-temperature synthesis reactions is attempted using a finite difference method. The model is described and its potential for estimating combustion wave velocities and temperature profiles is examined. Experimentally determined wave velocities on the self-propagating reaction of $3TiO_2 + 4Al + 3C \rightarrow 3TiC + 2Al_2O_3$ are compared with model results and are found to have good agreement.

1. INTRODUCTION

Conventional sintering techniques of producing ceramic materials generally require high furnace temperatures and relatively long processing times. In order to reduce the high energy requirements of the process a technique termed self-propagating high-temperature synthesis (SHS) has recently been attracting some interest [1, 2]. The basis of SHS is to use highly exothermic reactions to produce a wide variety of ceramic and intermetallic materials. When a SHS reaction is initiated in one area of a reactant mixture (*e.g.* with a resistively heated tungsten wire) there is sufficient heat release that the reaction becomes self-propagating. A combustion wave travels along the reactants converting them to the required products. The potential advantages are that the process requires little energy and the processing time is reduced to seconds rather than hours. In addition, SHS-produced materials are generally of high purity as the high temperatures attained by the product during passage of the combustion wave allows volatile impurities to be expelled [1].

Although there are many reports on the production of ceramic materials by various exothermic reactions, there is little work on modelling of the SHS process. Merzhanov [3] and Novozhilov [4] derived an analytical expression for the combustion wave velocity which was dependent on the activation energy (E) of the reaction. Hardt and Phung [5] developed a model whereby the rate of reaction was determined by the time required for reactant atoms in adjacent particles to diffuse through the product which formed between them. The wave velocity was therefore dependent on the activation energy for diffusion (Q) and diffusion coefficient (D). These models require knowledge of parameters which are not easily available or calculated.

A more recent approach has employed simple finite-different heat flow calculations to model the SHS process and has been used with some success to calculate wave velocities and parameter effects for the Ti + C → TiC reaction [6, 7, 8]. The basis of this paper is to examine if a finite difference method can estimate wave velocities and temperature profiles particularly for the reaction,

$$3TiO_2 + 4Al + 3C \rightarrow 3TiC + 2Al_2O_3 \qquad (1)$$

which produces a multiphase composite [9] with potential for high hardness and wear resistance.

2. MODEL FORMATION

A reaction becomes self-propagating when there is sufficient heat release by the reaction to ignite reactants ahead of the reaction zone. The basis of this model is to try and emulate this process using a finite difference method. Therefore, the temperature of the reaction zone (adiabatic combustion temperature) and the temperature at which the exothermic reaction begins (ignition temperature) must be known.

2.1 Adiabatic combustion temperature and ignition temperature

The adiabatic combustion temperature (T_{ad}) can be calculated by considering that the enthalpy of the reaction ($\Delta H_{r,298}$) heats up the products and that no energy is lost to the surroundings. From the definition of heat capacity of the product,

$$-\Delta H_{r,298} = \int_{298}^{2323} C_p(3TiC + 2Al_2O_3) \, dT + \Delta H_m(2Al_2O_3)$$

$$+ \int_{2323}^{T_{ad}} C_p(3TiC + 2Al_2O_{3(liq)}) \tag{2}$$

where $C_p(3TiC + 2Al_2O_3)$ is the specific heat of the product. The enthalpy of fusion of alumina, $\Delta H_m(2Al_2O_3)$, is included in the equation as T_{ad} is greater than the compounds melting point of 2323 K. From thermodynamic data the calculated value of T_{ad} is 2390 K.

The ignition temperature for this reaction was determined from DTA experiments and also by the process of thermal explosion whereby the reactants are gradually heated in a furnace until the reaction takes place throughout the whole sample rather than by a propagating combustion wave. The ignition temperature was determined to be approximately 900°C, as also observed by Abramovici [10].

2.2 Finite difference model

The finite difference model considers the reactants as a one dimensional series of cells which are initially at room temperature. The reaction is initiated by reacting the first two cells to the adiabatic combustion temperature to imitate rapid heating by an external heating source. Heat flow between cells heats up the third unreacted cell. Once this cell reaches the ignition temperature it is instantaneously converted to products at the adiabatic combustion temperature. The fourth unreacted cell then begins to heat up and the process begins again so that a combustion wave travels along the reactants.

Heat flow calculations can be carried out using an energy balance approach [11]. Consider the three adjacent cells (m − 1, m, and m + 1) in Figure 1,

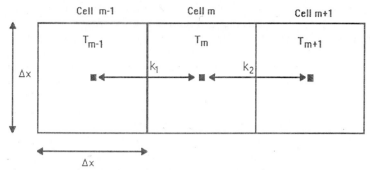

Figure 1. Three neighbouring cells in the one
dimensional array.

which are at temperatures T_{m-1}, T_m and T_{m+1} respectively. The explicit
finite difference equation is,

$$k_1 A \frac{[T_{m-1}(t) - T_m(t)]}{\Delta x} + k_2 A \frac{[T_{m+1}(t) - T_m(t)]}{\Delta x}$$
$$= \rho \cdot A \cdot \Delta x \cdot C_p \cdot \frac{[T_m(t + \Delta t) - T_m(t)]}{\Delta t} \quad (3)$$

where,

k_1 = thermal conductivity between cell $m - 1$ and m so $k_1 = 2 \cdot k_{m-1} \cdot k_m / (k_{m-1} + k_m)$.

k_2 = thermal conductivity between cell $m + 1$ and m(J s^{-1} m^{-1} K^{-1})

C_p = heat capacity of cell (J Kg^{-1} K^{-1})

ρ = density of cell (Kg m^{-3})

Δx = cell dimension (m)

Δt = time increment between calculations (s)

A = cross sectional area of cell (m^2)

$T_m(t)$ = temperature of cell m at time t(K)

$T_m(t + \Delta t)$ = temperature of cell m at time t + Δt(K)

Rearranging Equation 3 allows calculation of the temperature of a cell at
time t + Δt by knowledge of the temperature of the cells at time t.

$$T_m(t + \Delta t) - T_m(t)$$
$$= \frac{\Delta t}{(\Delta x)^2 \cdot C_p \cdot \rho} \{ k_1 \cdot [T_{m-1}(t) - T_m(t)] + k_2 \cdot [T_{m+1}(t) - T_m(t)] \} \quad (4)$$

The first cell is allowed to cool by convection and radiation so that the
equation for this cell is,

$$\varepsilon \sigma A [298^4 - T_1{}^4(t)] + hA [298 - T_1] + k_2 A \frac{[T_2(t) - T_1(t)]}{\Delta x}$$
$$= \rho \cdot A \cdot \Delta x \cdot C_p \frac{[T_1(t + \Delta t) - T_1(t)]}{2\Delta t} \quad (5)$$

where ε is the emissivity of the surface, σ is the Stefan Boltzmann constant and h is the heat convection coefficient.

A computer program calculated the temperatures of a row of 100 cells at time intervals Δt. The reactant cells had the thermophysical properties of the $3TiO_2 + 4Al + 3C$ powder mixture. Once a reactant cell reached the ignition temperature it was instantly converted to the adiabatic temperature and given the thermophysical properties of the $3TiC + 2Al_2O_3$ product. The cell size is given the approximate value of the particle size of the reactants used, which here was taken to be 15 μm (considering 10 μm Al particles surrounded by sufficient TiO_2 and C to cause complete reaction) and calculations are carried out at time intervals Δt which are small enough to allow calculations to be stable. Combustion wave velocity is measured by calculating the time travelled between cells and in addition data is stored in order to produce time-temperature histories of cells and also distance-temperature profiles.

2.3 Thermophysical properties

In order to model the SHS reaction one must have a knowledge of the values of C_p, ρ and k for the reactant powder mixture and the product.

2.3.1 Specific heat and density

The specific heat and theoretical density of the compounds are readily available in the literature [12, 13]. As the reaction is highly exothermic this implies that tighter bonds are being formed and thus the theoretical density of the products (4360 kg m^{-3}) is higher than the theoretical density of the reactants (3347 kg m^{-3}). However, in combustion experiments the volume of the sample does not change significantly during passage of the combustion wave. This therefore means that the actual density of the sample does not change but porosity is generated in the sample and this is accounted for. The initial reactant porosity is ~ 50% and so the product is ~ 60% porous. This porosity increase during synthesis is one of the major drawbacks of the SHS process and a densification process is usually required [14].

2.3.2 Thermal conductivity of the reactant powder mixture

Although there are reference tables available on the thermal conductivity of solid materials there is little data on the thermal conductivity of powders and powder mixtures. A model developed by Luikov et al. [15] was used which has been able to predict the thermal conductivity of a variety of metal and ceramic powders. The basis of the model is to calculate the effective thermal conductivity of a powder by estimating the contributions of three main conduction mechanisms which are conduction through the gas between the powder particles, radiation conduction between particles and contact thermal conduction between particles. The model requires values of the density, particle size, Elastic modulus (to calculate contact area) and the conductivity of air and solid material. Table 1 lists the calculated values of k for the three powder compounds which compare well with values obtained from literature (although these have slight differences is density and are of unknown particle size).

Table 1. Comparison of calculated thermal conductivity of the powders used (50% theoretical density) with literature values

Powder (particle size)	Calculated k (W m⁻¹ K⁻¹)	Literature k (W m¹ K⁻¹)
Carbon black (0·02 μm)	0·015	0·020 [16]
TiO₂ (0·7 μm)	0·079	0·068 [17]
Aluminium (10 μm)	0·291	0·256 [17]

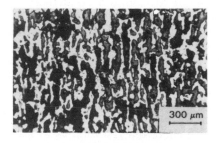 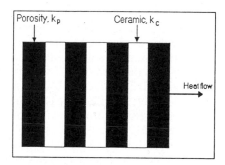

**Figure 2. (a) Optical micrograph of TiC-Al₂O₃ formed by SHS.
(b) Idealised microstructure as an alternating layer of product and porosity.**

The value, k, of the powder mixture was determined by a three phase rule of mixtures [18] assuming that the carbon black is a continuous phase as it has such a small particle size. The thermal conductivity of the powder mixture was calculated to be 0·081 Wm⁻¹ K⁻¹. The variation of thermal conductivity with temperature to the ignition temperature was neglected in the model as during SHS reactions gases and volatile species are violently expelled. As conduction in powders in primarily by conduction in the gas phase the accurate determination of k in the reaction zone region would be extremely difficult.

2.3.3 Thermal conductivity of the product

Figure 2(a) reveals the microstructure of the TiC-Al₂O₃ composite formed by the advancing combustion wave. The dark area is porosity and the light area is the composite. The microstructure can be imagined as alternating layers of porosity and ceramic as shown in Figure 2(b). If we consider heat flow in the direction shown then heat conduction is dominated by the poorer conductor. The overall thermal conductivity (k_m) is,

$$k_m = \frac{k_c \cdot k_p}{(V_p \cdot k_c + V_c \cdot k_p)} \tag{6}$$

where k_c and k_p are the thermal conductivity of the ceramic and pore respectively and the V_c and V_p are the volume fractions of the ceramic and porosity. As $k_c >> k_p$ then $k_m \sim k_p/V_p$.

The value of k_p of a pore is a combination of the conductivity of the air in the pore and the radiation across pores [19], such that,

$$k_p = k_{air}(T) + 4d_p\sigma\varepsilon T^3 \qquad (7)$$

where $k_{air}(T)$ is the conductivity of air at temperature T and d_p is the pore size.

3. MODEL RESULTS AND COMPARISON WITH EXPERIMENTAL DATA

Figure 3 shows a typical time-temperature profile for a single cell. In the region AB the reactants are gradually heated up by the advancing combustion wave with the aluminium melting. On reaching the ignition point there is a rapid temperature increase as the reactants are converted to products (BC). The cell begins to lose heat to its neighbouring cell and thus cools rapidly along CD with the alumina solidifying. The reactants in the next cell then begin to heat up to the ignition point. The periodic rise and fall of the cell temperature in the region DE is caused by the back flow of heat from the advancing combustion wave. Similar irregularities have been observed in experimentally measured time-temperature profiles but it is uncertain if this is due to the same mechanism [20].

The combustion wave velocity was determined from the time histories of two cells a known distance apart as shown in Figure 4. Table 2 shows the agreement between the calculated and experimentally determined combustion velocity and temperature. The wave velocity was determined experimentally from a series of photographs of the combustion wave taken at a known time interval and the combustion temperature was measured via optical pyrometer.

In addition to time-temperature histories the model can produce a time-distance profile of the advancing combustion wave at a time t as seen in Figure 5, with the direction of propagation shown. The profile reveals the high temperature gradient between the reactants and products and the cooling of the product after passage of the wave. Increasing the thermal conductivity of the powder mixture lowers the temperature gradient but causes rapid cooling

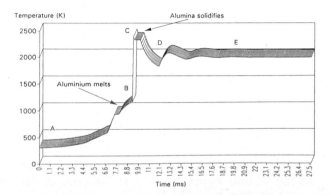

Figure 3. Time-temperature profile of a single cell for the finite difference model.

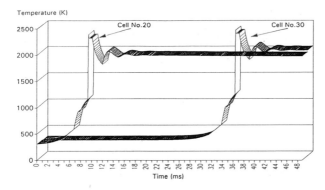

Figure 4. Calculation of combustion velocity is achieved from the time-temperature histories of two cells of known distance apart.

Table 2. Comparison of calculated and measured combustion temperature and wave velocity

	Combustion temperature (K)	*Wave velocity (mm s⁻¹)*
Model	2390	5·8
Measured	2206 ± 100	3·2 ± 0·2

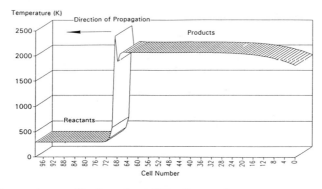

Figure 5. Temperature profile throughout 100 cells at a time t showing the high temperature gradient between the reactants and products and cooling of the product after passage of the combustion wave.

of the reaction zone and the combustion wave is extinguished. This has been noted by Rice *et al.* on compacts of high density which are of high thermal conductivity [21].

3.1 Model results for other SHS reactions

In order to evaluate if the model is applicable to other SHS reactions, model wave velocities were compared to literature results. The ignition temperatures were taken from thermocouple measurements of time-temperature profiles

C. R. Bowen and B. Derby

Table 3. Model results and combustion wave measurements from literature

Reaction	T_{ad} (K)	T_{ig} (K)	Particle (μm)	Cell size (μm)	V_{meas} (mm s^{-1})	V_{model} (mm s^{-1})
Ti + C $\rightarrow$ TiC* [6]	3210	1373	Ti (44), C (2)	50	8	9
Ti + 2B $\rightarrow$ TiB$_2$* [5]	3200	1573	Ti (10), B (0·5)	10	68	64
Ti + 2B $\rightarrow$ TiB$_2$* [5]	3200	1573	Ti (60), B (0·5)	60	48	11

*Solidifiction neglected

[22, 23] and the thermal conductivities of the powders are experimentally measured values of k. Table 3 lists the reactions and the results obtained. For the Ti + C $\rightarrow$ TiC reaction the calculated velocity agrees well with the experimentally determined value and also the velocity of 10 mm s^{-1} determined with the finite difference modelling of Advani *et al.* [8] (although different ignition temperatures were used).

For the $Ti + 2B \rightarrow TiB_2$ reaction, Table 3 shows that increasing the cell size of the model causes a decrease in the wave velocity in the same way as increasing the particle size. The reduction of wave velocity with particle size is, however, not a heat transfer effect. Smaller particles have greater surface area to react and there are less diffusion barriers to slow down the reaction (Hardt and Phungs model considers this [5]). The effect of wave velocity on cell size can be realised if one considers the model cells getting so small that it will take an infinitesimally small time for an unreacted cell to reach the ignition temperature and thus the wave velocity increases dramatically. From Equation 4 if one considers the time Δt for an equal change in temperature of two cells of size Δx_1 and Δx_2 then $\Delta t_1 / \Delta t_2 = [\Delta x_1 / \Delta x_2]^2$. As combustion wave velocity (V) is $\Delta x / \Delta t$ then the variation of velocity with cell size is $V_1 / V_2 = \Delta x_2 / \Delta x_1$.

4. CONCLUSION

The finite difference model is able to predict combustion wave velocities and produce temperature profiles by considering the reaction rate being limited by heat transfer only. Time temperature profiles reveal a periodic fluctuation in temperature directly after the combustion reaction, caused by back heating of the advanced combustion wave. Distance-temperature profiles show the large temperature gradient between reactants and products which is required to sustain the reaction. Increasing the conductivity of the powder reduces the temperature gradient but causes rapid cooling of the reaction zone and results in the combustion wave being extinguished.

The model of Luikov et al. enables the thermal conductivity of powder mixtures to be calculated, allowing modelling of a variety of reactions without the need to measure k experimentally. As density is a parameter in the Luikov model the effect of reactant density on the $3TiO_2 + 4Al + 3C$ reaction (and other reactions) may be analysed.

Model reaction velocities have been shown to decrease with increasing cell size. However, the mechanism is different to the decrease in wave velocity with particle size. In addition, the reaction zones can sometimes be of the order of millimetres [22] compared with the cell size of micrometres so that a representative value of model cell size is a difficult choice.

ACKNOWLEDGMENT

The authors gratefully acknowledge financial support for this work from SERC and the Cookson Technology Centre. In addition I would like to thank Dr. J. Hunt for his help on heat transfer.

REFERENCES

1. MUNIR, Z. A. & ANSELMI-TAMBURINI, U., *Mat. Sci. Rep., 3,* 277, (1989).
2. YI, H. C. & MOORE, J. J., *J. Mat. Sci., 25,* 1159, (1990).
3. MERZHANOV, A. G., *Dokl. Akad. Nauk SSSR, 233,* 1130, (1977).
4. NOVOZHILOV, B. V., *Dokl. Akad. Nauk SSSR, 144,* 1328, (1962).
5. HARDT, A. P. & PHUNG, P. V., *Combust. Flame, 21,* 77, (1973).

6. KOTTKE, T., KECSKES, L. J. & NILLER, A., International Symposium on Combustion Synthesis of High Temperature Materials, San Fransisco, October, (1988).

7. KOTTKE, T. & NILLER, A., *Thermal Conductivity Effects on SHS Reactions,* Technical Report BRL-TR-2889, Ballistics Research Laboratory, Aberdeen Proving Ground, Maryland, (1988).

8. ADVANI, A. H., THADHANI, N. N., GREBE, H. A., HEAPS, R., COFFIN, C. & KOTTKE, T., *Scripta Met.,* **25,** 1447, (1991).

9. BOWEN, C. R., HULSMANN, S. & DERBY, B., presented at The Second European Ceramic Society Conference, Augsburg, September, (1991).

10. ABRAMOVICI, R., *Mat. Sci. and Eng.,* **71,** 313, (1985).

11. INCROPERA, F. P. & DE WITT, D. P., Introduction to Heat Transfer (2nd ed.), publ. John Wiley and Sons, New York, (1990).

12. KUBASHEWSKI, O. & ALCOCK, C. B., Metallurgical Thermochemistry, Pergammon Press, New York, (1979).

13. MORRELL, R., Handbook of Properties of Technical and Engineering Ceramics, Her Majesty's Stationery Office, London, (1985).

14. RICE, R. W. & McDONOUGH, W. J., *J. Am. Ceram. Soc.,* **68,** C122, (1985).

15. LUIKOV, A. V., SHASHKOV, A. G., VASILIEV, L. L. & FRAIMAN, YU. E., *Int. J. Heat Mass Transfer,* **11,** 117, (1968).

16. ROGERS, D. B. & WILLIAMSON, J. W., *Thermal Conductivity,* **15,** 317, (1978).

17. MEDVEDEV, N. N., *J. Eng. Phys.,* **14,** 176, (1968).

18. BRAILSFORD, A. D. & MAJOR, K. G., *Brit. J. Appl. Phys.,* **15,** 313, (1964).

19. LOEB, A. L., *J. Am. Ceram. Soc.,* **37,** 96, (1954).

20. DUNMEAD, S. D., MUNIR, Z. A., HOLT, J. B. & KINGMAN, D. D., *J. Mat. Sci.,* **26,** 2410, (1991).

21. RICE, R. W., RICHARDSON, G. Y., KUNETZ, J., SCHROETER, T. & McDONOUGH, W. J., *Adv. Ceram. Mat.,* **2,** 222, (1987).

22. DEEVI, S. C., *J. Mat. Sci.,* **26,** 2662, (1991).

23. DUNMEAD, S. D. & HOLT, J. B., Proceedings of the DARPA/Army Symposium on SHS, Daytona Beach, Florida, October, (1985).

Microwave Sintering of Advanced Ceramics

F. C. R. WROE and J. SAMUELS

EA Technology, Capenhurst, Chester, CH1 6ES

ABSTRACT

EA Technology is involved with the microwave processing of ceramics, particularly advanced ceramics, and has extensive experience in the design and operation of a number of dedicated microwave furnaces. A dilatometer for the continuous in-situ measurement of the densification process in both the conventional and microwave environments has also been designed and is currently operational.

This paper will draw on that experience to discuss the results of a series of comparative experiments, in microwave and conventional furnaces, which have shown an enhancement of the sintering process, as measured by end-point densities and through the use of the dilatometer. Results will be presented for both oxide-based and non-oxide-based ceramics.

1. INTRODUCTION

Microwave sintering is a novel electrical process technique for ceramic materials which differs fundamentally from the current conventional processes by providing volumetric heating. As a result of this feature several technical and economic advantages have been cited for this technique. These include [1–3], more rapid and uniform heating, improved microstructural properties, shorter furnace response times and enhanced energy efficieny.

Furthermore, there have been claims made in the literature that the enhanced densification observed has been due to a reduction in the activation energy for sintering, although the micromechanisms for this are not yet understood and it is this which is usually described as the "microwave effect."

There are two main physical loss mechanisms by which microwaves interact with ceramics, resulting in internal heating. These are the flow of conductive currents (in particular ionic conduction) and dipolar reorientation. Mathematically, both of these losses may be included in an effective dielectric loss factor.

$$\varepsilon_e'' = \varepsilon' + \frac{\sigma}{\omega\varepsilon_o}$$

$$\text{dipolar losses} \qquad \text{conductive losses}$$

where ε'' dielectric loss factor
  σ conductivity
  ε_o permittivity of free space
  ω frequency

An alternative expression of the losses which is frequently used is the effective loss tangent $\tan\delta_e$:

$$\tan\delta_e = \frac{\varepsilon_o''}{\varepsilon'} = \frac{\varepsilon''}{\varepsilon'} + \frac{\sigma}{\omega\varepsilon'\varepsilon_o}$$

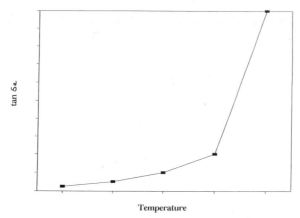

Figure 1. Typical variation of tan δe with
temperature for a low loss ceramic.

where ε' relative dielectric constant

The power deposited in the ceramic dielectric is then given by

$$P = \omega\varepsilon_0\varepsilon' \tan \delta_e |E|^2$$

where   E is the applied electric field.

Figure 1 shows how tan δ_e varies in a low loss ceramic, such as alumina, for which there is some data available in the literature. At room temperature tan δ_e is very low and the ceramic is essentially transparent to microwave radiation. Above some critical temperature, which appears to vary according to the ceramic used, tan δ_e rises rapidly, resulting in more effective heating. This behaviour has been attributed by some to an increase in the ionic conductivity with temperature. The heating of the ceramic from room temperature to the point where it becomes lossy often requires some form of external pre-heating. Reports have been made in the literature of the use of an external susceptor, which is a material with high losses at low temperatures, and/or the addition of a lossy material to the green body which may or may not be removed during processing.

The rest of the paper will be concerned with a description of the facilities we have at EA Technology, a presentation of some of the results obtained for both oxide and non-oxide-based materials, together with a brief description of some of the work on the sintering dilatometer which is starting to provide data on the reasons for the densification enhancement observed.

2. THE FACILITIES AT EA TECHNOLOGY

Over the past two years we have established a substantial facility for the microwave sintering of ceramics. At present we have two microwave furnace applicators, one with a hot zone of $200 \times 200 \times 200$ mm and one with a hot zone of $250 \times 250 \times 300$ mm. One of them, the larger, also has a controlled atmosphere capability for inert gases, and both are capable of 1750–1800°C.

The furnaces can be attached to a variety of power supplies, all at 2,450 MHz, and varying in power levels from 1·2 to 5 KW. Emphasis has been placed on the power and temperature control aspects which are very important for the ultimate exploitation of this process. The power to the furnace is controlled through a Eurotherm 818P programmer, which responds to input from the temperature measurement system, an Accufiber system from the USA. We use three types of temperature measurement devices, a black body probe, a light pipe and various pyrometers, depending on the precise requirement of the work. In summary we attempt to achieve total integrated process control throughout all stages of the firing cycle.

Much emphasis has been placed in the literature regarding the uniform heating profile achieved when using microwave energy for sintering. In certain circumstances this is true, but as the sample size increases, there is the possibility that an inverse profile could develop with a hot spot forming in the centre of the specimen, due to the heat losses being limited to those occurring at the surface of the specimen. This would obviously be undesirable, and lead to thermal stresses developing in the sample. A major part of our work, therefore, is the development of a uniform heating profile throughout the sample, which we can now do successfully, using a variety of techniques, some of which are being patented.

Furthermore, we have designed and constructed a sintering dilatometer [4], capable of operating in both conventional and microwave heated furnaces, in order to monitor sintering rates continuously during firing. This dilatometer is constructed from recrystallised alumina tubing, and utilizes a linear variable differential transformer (LVDT) transducer to detect displacements with a resolution of 0·1 microns. The microwave furnace is capable of reaching over 1700°C at a ramp rate of 30°C/min in air. Another is being designed at present which will be capable of the same temperatures and heating rates, but with a controlled atmosphere. A novel susceptor design is used, allowing repeated rapid pre-heating from room temperature. Data from the transducer and temperature measurement system is continually logged during the heating cycle and then processed using a spreadsheet type programme in order to calculate both the linear shrinkage and the sintering rate.

3. EXPERIMENTAL PROCEDURE

The following materials were used in the experimental work, the results of which will be reported in later sections.

3.1 Oxide materials

(a) Alcoa aluminium oxide grade A1000SG.
(b) Zircar Fibrous Ceramics ZYP 12 wt% yttria stabilised zirconium oxide powder.

Both powders have submicron grain sizes, with the zirconia having a very fine crystalline size of 0·02–0·03 microns.

3.2 Non-oxide materials

(a) Starck aluminium nitride grade C.
(b) Starck silicon nitride grade LC12N.
(c) Starck yttrium oxide, fine grade.
(d) Alcoa aluminium oxide grade A1000SG.
(e) Tosoh zirconium oxide grade TZ3Y.

The oxides mentioned above were used as either sintering additives, or, in the case of the zirconia, to attempt to enhance the composition's microwave susceptibility. The samples were prepared by wet-milling the powders in appropriate quantities, water was used for the oxide work, iso-propanol for the non-oxide. The oxide work had 2–3% polyethyleneglycol (PEG) added, the non-oxide had 1·5% PEG and 1·5% polyvinylbutryal added. In each case the powders were dried, crushed and sieved prior to uniaxial pressing using a 19 mm tool steel die at a pressure of 96 MPa. Samples for the dilatometer work were pressed at the same pressure in a 12 mm die. The binder(s) were burnt out of the compacts in a muffle furnace using a slow heating cycle prior to any firing cycle, and the resulting compacts had a green density of 55 to 60%.

Sintering trials were carried out using a variety of furnace set-ups. Most of the oxide work was done in a 700 W, 2·45 GHz domestic microwave oven. The non-oxide work was carried out in an atmosphere-controlled 5 KW, 2·45 GHz, custom-built furnace applicator. The non-oxide sintering utilized powder-bed technology, the powder-bed being the same composition as the sample, but with 20 weight percent boron nitride added. The dilatometer experiments were performed in a 1·4 KW, 2·45 GHz commercial oven. Each of the experiments used a susceptor, the susceptor for the oxide work consisted of zirconia fibre loosely compacted in an alumina crucible. The non-oxide, like the dilatometer work mentioned earlier, employed a susceptor of a novel design, which permits repeated, rapid heating from room temperature without the use of any conventional pre-heat which is required for the zirconia fibre susceptor described above.

Initial work involved the manual switching of the microwave power, but subsequently a Eurotherm 818P programmable controller, linked to the Accufiber system was used. A predetermined heating profile was followed in all cases, up to the maximum sintering temperature. For comparison, samples were heated conventionally in a rapid heating Super Kanthal furnace for the oxide work, a tube furnace for the non-oxide work and a split Super Kanthal furnace for the dilatometer experiments. The heating profile used was principally determined by the maximum heating rate of the conventional furnaces, and not the microwave furnaces.

Following sintering, the final densities were measured by the Archimedes technique, with mercury as the immersion medium, or by physical measurement. X-ray Diffraction (XRD) analysis was performed on selected samples, and microstructural observations were carried out on the scanning electron microscope (SEM).

4. RESULTS AND DISCUSSION

4.1 Oxide compositions

Results are presented for the following three oxide compositions, which are in order of increasing microwave susceptibility.

(i) Alumina.
(ii) Alumina + 50 wt% yttria stabilised zirconia (YSZ).
(iii) YSZ.

Figure 2 shows the variation of final density (expressed as a percentage of theoretical density) with sintering temperature for both the microwave and conventional sintering of alumina. All the oxide compositions presented today had no hold time at temperature. It can be seen that there is a microwave enhancement of the end point density when compared to their conventional equivalent and that this is most evident at low temperatures and low densities.

The alumina samples reach a specified end point density approximately 50°C below that of the conventionally fired samples. Kimrey et al. [5] have also observed enhanced microwave sintering of alumina. In their system (28 GHz, 200 KW, hard vacuum) a temperature difference of 300–400°C was measured.

Figure 3 shows the variation with final density for both the conventional and microwave sintered alumina with 50 w/o yttria stabilised zirconia. This once more shows an enhancement in final density for the microwave-fired specimens of approximately the same magnitude as in the 100% alumina samples. Both the microwave and conventionally fired samples show an increase in density compared to alumina. This indicates that YSZ enhances the sintering of alumina. Once more Kimrey et al. [6] have also demonstrated that the sintering of alumina-zirconia composites is enhanced by the use of microwaves.

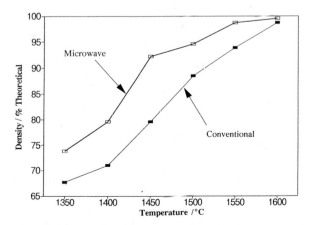

**Figure 2. Final   density   comparison   —
            alumina.**

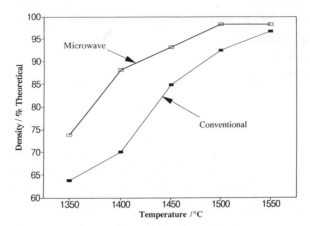

Figure 3. Final density comparison Al₂O₃-50 wt% YSZ.

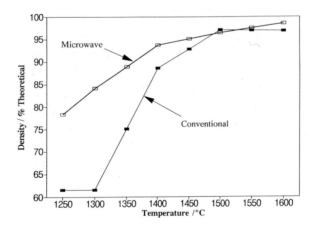

Figure 4. Final density comparison — YSZ.

Finally, Figure 4 shows the data for 100% YSZ, which is the most microwave susceptible of all the oxide compositions presented here. The enhancement of the microwave sintering is most pronounced, especially below 1500°C and the difference in equivalent densities here is as much as 100°C between the microwave and conventionally fired samples.

Microstructural evaluation of the samples revealed that the microwave sintered samples have slightly larger grain sizes, for a specified temperature which could be expected from their heigher densities. For the zirconia-containing alumina the micrographs revealed a uniform distribution of the zirconia throughout the material.

4.2 Non-oxide compositions

Results will be presented here for four of the non-oxide compositions studied to date. The compositions are as follows:

Series A: Si_3N_4 with 4 wt% Al_2O_3 and 6 wt% Y_2O_3

Series B: β'-sialon (z = 1) with 6 wt% Y_2O_3

Series C: Si_3N_4 with 4 wt% Al_2O_3, 6 wt% Y_2O_3 and 10 wt% ZrO_2 (containing 3 mol percent yttria).

Series D: β'-sialon with 6 wt% Y_2O_3 and 10 wt% ZrO_2 (containing 3 mol percent yttria).

Figures 5 to 8 show the results obtained for Series A to D; These show that these systems also illustrate the microwave effect of enhanced sintering at

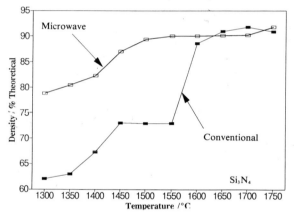

Figure 5. Final density comparison — Series A.

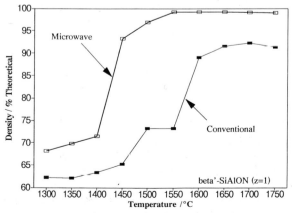

Figure 6. Final density comparison — Series B.

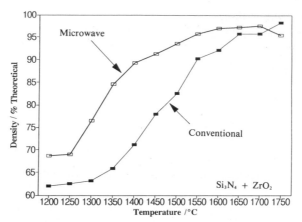

Figure 7. Final density comparison — Series C.

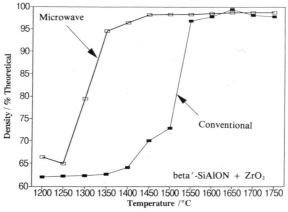

Figure 8. Final density comparison — Series D.

lower densities and temperatures. The microwave profiles are very similar to the conventional profiles, but the microwave sintered samples reach a specified density at a temperature between approximately 100 to 200°C lower, depending on the system studied.

A limited amount of information is published in the literature on the sintering of silicon-nitride-based materials, with a couple of papers recently published by Tiegs *et al.* from ORNL [7]. Although they did not have atmosphere control on their applicator design, they buried their samples in a powder-bed. They did not see the dramatic sintering enhancement observed in their oxide work, with a maximum temperature difference, due to microwave enhancement of 50°C at 1750°C. No results have been published at the lower temperatures, but their results show that they observed no enhancement below

1700°C which is in contradiction to the results presented here today. However, they were using a different silicon nitride source and a slightly different liquid phase composition. The explanation given for the lack of a dramatic reduction in sintering temperatures is that the liquid phase must still be present for microwave sintering, as well as conventional sintering, to occur. The difference in the results obtained at EATL, could be explained by the fact that we have shown that liquid phases are formed at a lower temperature in microwave sintering than is observed conventionally, or alternatively, that when liquid phases begin to be formed, the microwave energy couples into them very effectively, enhancing their formation and thereby reducing the sintering temperature. At present there is not sufficient evidence to explain the difference in results, except to say that ORNL were using different starting materials, liquid phase compositions and processing techniques.

The actual end-density profile is dependent on whether it is a silicon nitride-based sample or a β'-sialon sample which is to be expected. The zirconia containing microwave processed samples do show an enhancement in the end-point density, but this is also observed in the conventionally fired samples. There doesn't appear to be any further enhancement due to the presence of zirconia in the microwave field as Series C appears to have a similar enhancement as Series A, and Series D is perhaps slightly enhanced due to the presence of zirconia over Series B. Therefore, the main effect in these systems is from the "microwave effect" and not the addition of zirconia. This confirms the results shown in the oxide work, that a substantial amount of zirconia is required before a significant increase in the "microwave effect" is observed.

Finally, Figures 9 to 12 show that it is not only the end-point density of silicon-nitride-based samples which is enhanced by microwave sintering, but also the conversion rate of α silicon nitride to β silicon nitride or β'-SiAlON is also enhanced. This was established from XRD measurement by averaging the

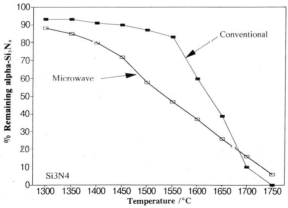

**Figure 9. Alpha-Si$_3$N$_4$ conversion —
Series A.**

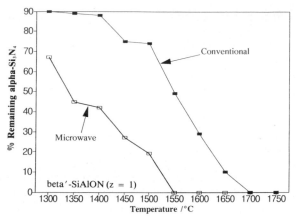

**Figure 10. Alpha-Si₃N₄ conversion —
Series B.**

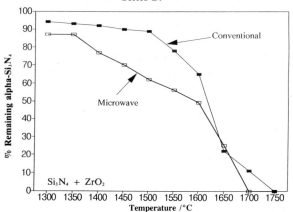

**Figure 11. Alpha-Si₃N₄ conversion —
Series C.**

peak intensities for both alpha and beta silicon nitride (β'-sialon). The peaks used were hkl = 201, 102 and 210 for alpha and hkl = 200, 210 and 102 for beta (or beta').

This is to a certain extent to be expected as the solution-reprecipitation mechanism is believed to have two simultaneous consequences, densification and phase transformation. However, recent work by Yuan *et al.* [8] both theoretically and experimentally has shown that the phase transformation is temperature-dependent, and significantly different from the densification process especially at lower temperatures, which means that calculations can be carried out on the rate of α to β conversion independent of the density.

Following on from this calculations are being carried out which calculate the activation energy for the rate of α to β conversion. Initial results show that in a microwave environment the activation energy for this conversion is reduced.

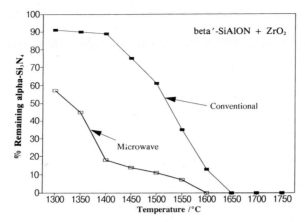

**Figure 12. Alpha-Si$_3$N$_4$ conversion —
Series D.**

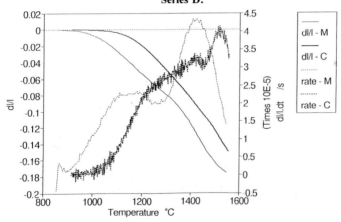

Figure 13. Microwave and conventional sintering profiles for alumina.

4.3 Results and discussion for the oxide compositions in the sintering dilatometer

Figures 13 to 15 show the results of the equivalent sintering dilatometer tests. Each graph compares the linear shrinkage (dl/l) and the sintering rates (dl/l · dt) observed in both the microwave and conventional furnaces. Each of the experiments was performed using a 5°C/min temperature rise. Again a significant amount of the enhancement of the sintering is observed.

For alumina the conventional sintering starts at about 1100°C, and the rate of sintering steadily increases to reach a maximum at about 1500°C. The microwave sintering starts sooner at about 900°C, reaches a first maxima at 1000°C, with the second maxima at 1500°C.

For the alumina/50 weight% zirconia composition, again the sintering for the microwave sample starts earlier at 900°C, with a small first maxima at 1000°C, and the peak sintering rate occurs at about 1300°C. The conventional sintering rates appear in this instance to be very similar, but the maxima are displaced by about 100°C, with the peak sintering rate occurring at approximately 1400°C.

Finally for the yttria stabilised zirconia, the displacement of the sintering rate by 100°C is once more observed, with both the microwave and conventional samples displaying similar sintering rate profiles.

Wang and Raj [9], using an analogy to the study of phase transformation and thermal desorption, suggested that the position and magnitude of the peak

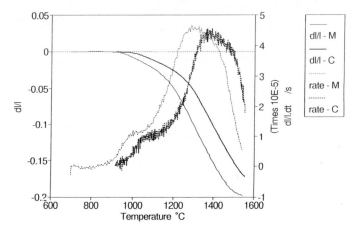

Figure 14. Microwave and conventional sintering profiles for alumina/50 wt% zirconia.

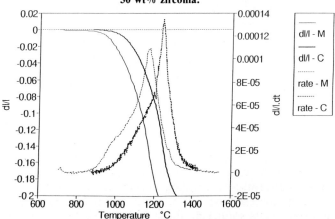

Figure 15. Microwave and conventional sintering profiles for zirconia.

sintering rate may lead to an insight into the activation energy for the sintering process. Their analysis together with some experimental results, indicates that peak sintering rates at higher temperatures are obtained with larger activation energies, with the temperature being the dominant factor. Following this argument, data from the tests would suggest a clear reduction of the activation energy for sintering during microwave processing, as has been claimed by Kimrey *et al.* [5]. Experiments are in progress to directly measure the activation energies using the dilatometer, and initial results suggest that it is more complicated than originally thought.

Early attempts to elucidate the nature of the microwave effect for non-oxide materials in the dilatometer without atmosphere control showed that the sintering rate was being significantly affected by the rate of oxidation and weight loss. Therefore, experiments will be performed in the non-oxide sintering dilatometer to attempt to establish the sintering enhancement mechanism, as soon as it has been commissioned with effective atmosphere control.

5. CONCLUSIONS

A comparative study of the microwave and conventional sintering of a series of advanced ceramics, both oxide and non-oxide, has established that microwave heating does enhance both the densification process and the rate of conversion of alpha to beta silicon nitride. This paper has also hopefully shown the importance of careful design in the microwave furnace applicators, in order, not only to optimise the microwave effect, but also its suitability for the system being studied.

REFERENCES

1. KATZ, J. D., BLAKE, R. D. & SCHERER, C. P., *Int. J. Refract. Met. & Hard Mat.*, 175, (1989).
2. BINNER, J. P. G., in *Proc. IEE Colloquium on "Industrial Uses of Microwaves,"* London, UK, (Digest No. 109, 2/1), (1990).
3. TINGA, W. R., *Mat. Res. Symp. Proc.*, **124**, 33, (1988).
4. SAMUELS, J. & BRANDON, J. R., to be published in *J. Mat. Sci.*
5. KIMREY, H. D., JANNEY, M. A. & FERBER, M. K., *Ceramic Technology Newsletter No.*, **20**, 3, (1989).
6. KIMREY, H. D., KIGGANS, J. O., JANNEY, M. A. & BEATTY, R. L., *Mat. Res. Symp. Proc.*, **189**, 243, (1991).
7. TIEGS, T. N., KIGGANS, J. O. & KIMREY, H. D., *Mat. Res. Symp. Proc.*, **189**, 267, (1991).
8. YUAN, L., YAN, D. & MAO, Z., *Chinese Sci. Bull.*, **35**, 914, (1990).
9. WANG, J. & RAJ, R. J., *Am. Ceram. Soc.*, **73**, 1172, (1990).

Applications from Temperature Profile Control During Microwave Sintering

J. G. P. BINNER* and T. E. CROSS†

*Department of Materials Engineering and Materials Design,
†Department of Electrical and Electronic Engineering, University of Nottingham,
University Park, Nottingham, NG7 2RD

ABSTRACT

The use of 2·45 GHz microwave radiation is being investigated for the sintering of a range of advanced ceramic materials in a number of separate research programmes at the University of Nottingham. This paper selects results from the projects which illustrate clearly the differences between microwave and conventional radiant sintering of ceramics and which indicate some of the advantages microwaves have to offer.

1. INTRODUCTION

The application of microwave energy to the processing of ceramic materials dates back at least three to four decades, however, for most of this period it has been uncoordinated and slow. There are many reasons for this, perhaps the major one being the user-supplier relationship in that each was largely unaware of the details and limitations of equipment and processes respectively [1]. This problem is now being specifically addressed with more time being allocated to clearly defining the requirements. Nevertheless, the diverse nature of materials which come under the heading of ceramics, and their fast development in recent years, have caused their own difficulties in fully exploiting the potential offered by the use of microwaves as an alternative energy source for ceramic processing. Table 1 summarises some of the potential applications of microwaves, the list is not intended to be definitive but to indicate topics where current research activity is ongoing around the world.

The greatest amount of effort with respect to research performed in the field of microwave-assisted ceramic processing is concerned with firing or sintering [1]. A wide variety of material systems have been investigated together with a number of different approaches. These latter have involved the use of both multi-mode and single-mode applicators. The former are more readily available and generally provide more uniform power dissipation, whilst the latter offer much more precise knowledge of the microwave field distribution and the ability to introduce significantly higher power loadings. Since the interaction between microwaves and materials is strongly materials property-dependent, and since many investigators have chosen to work with ceramics which have intrinsically low dielectric losses at low temperatures, three further approaches have involved the use of coupling aids, susceptors and high microwave frequencies. A discussion of the merits of these approaches has been provided elsewhere [2].

Table 1. Applications for microwave energy in ceramic processing

Synthesis	Synthesis of a range of non-oxide ceramic powders is being studied including silicon carbide and titanium carbide.
Drying	The majority of current work is focused on traditional ceramics, refractories and moulds for investment casting.
Slip casting	Microwaves have been shown to increase the rate of casting; also drying of moulds and ware.
Calcination	The heat treatment of precursors to yield the desired phase. Research to date has focused on superconducting ceramics and certain electroceramics, however little success has been reported.
Sintering	The greatest amount of work is being done in this field. A wide range of ceramic and ceramic matrix composite systems are currently being studied.
Joining	Attempts are being made to join both oxides and non-oxides with significant early success.
Plasma assisted sintering	The use of microwaves to generate plasmas to assist in oxide sintering. Plasma sintering is a research field in its own right, however some indication has been obtained that microwave created plasmas can provide superior results.
CVD	The use of microwaves to decompose gaseous species which recombine to form thin films or powders. CVD is a research field in its own right.
Other applications	Includes process control, clinkering of cement products, heating of optical fibre preforms, removal of organic binders from injection moulded ware, etc.

2. DIELECTRIC LOSS MECHANISMS IN CERAMICS

Dielectric losses can be formally described by considering the dielectric constant to be a complex number of the form [1]:

$$\varepsilon = \varepsilon_0(\varepsilon' - j\varepsilon'') = \varepsilon_0\varepsilon'(1 - j\tan\delta)$$

where the dielectric constant, ε', is related to electrical polarisability and the loss factor, ε'', determines energy dissipation. Tan δ is the ratio of dielectric constant to loss factor and is known as the loss tangent. The value measured at the appropriate frequency can provide a useful indication of the type of interaction a material will undergo in a microwave field (Figure 1). A number of ceramics have a loss tangent lower than about 0·01 and can be considered to be transparent to microwaves (Figure 1(a)). Highly conductive materials such as metals on the other hand, are opaque to microwaves (Figure 1(b)) resulting in almost total reflection. This is the well known basis behind radar detection. The vast majority of ceramic materials, however, fall between these two extreme types of behaviour and absorb microwaves to a lesser or greater degree (Figure 1(c)).

There are primarily two physical mechanisms through which energy can be transferred from microwave frequency radiation to a ceramic material [1]. Permanent dipoles exist in some ceramics and these tend to reorientate under

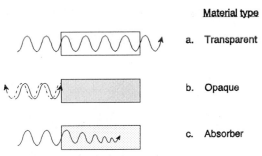

Material type

a. Transparent

b. Opaque

c. Absorber

Figure 1. Interaction of microwaves with materials (adapted from Ref. 3).

the influence of a microwave electric field, especially in the frequency range 1–10 GHz. The origin of the heating effect lies in the inability of the polarization to follow the extremely rapid reversals of the electric field. At such high frequencies, the resulting polarization phasor P lags the applied electric field ensuring that the resulting current density has a component which is in phase with the electric field and therefore power is dissipated in the ceramic. This mechanism is equally applicable to the heating of water where the permanent dipoles are the water molecules themselves. This led to the first experiments with microwave ceramic processing being associated with drying where the ability of the ceramic to couple with microwaves is of secondary importance.

The second mechanism involves the flow of conductive currents, resulting in an ohmic type of loss mechanism where the ceramic conductivity plays a prominent role. This mechanism is particularly prevalent at radio frequencies but can also occur at the higher microwave frequencies in semiconductor ceramics. Examples include silicon carbide, in which eddy currents are induced, and ceramics containing ionic impurities, where the conductive currents arise due to the movement of ionic constituents. Both the dipolar and ohmic heating mechanisms lead to volumetric power deposition and are capable of resulting in sufficient power being deposited to raise the temperature of many ceramics to that required for processes such as sintering (typically 1400–2000°C) provided heat losses to the surrounding medium can be restricted.

3. MICROWAVE SINTERING

Microwave heating is fundamentally different from more conventional radiant element techniques. Figure 2 illustrates a typical power deposition profile in a microwave irradiated dielectric, power dissipation following the exponential law:

$$P = P_t e^{-2\alpha x}$$

where P_t is the power transmitted through the surface in the x-direction and α is the attenuation constant. Since power is only deposited into the ceramic, the surrounding air remains cooler than the body, which can result in the creation of an inverse temperature profile with time (Figure 3), *i.e.* a hotter interior

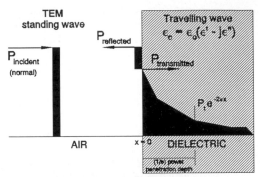

Figure 2. Power deposition profile in a dielectric medium.

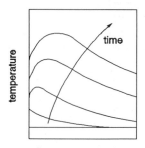

distance into specimen

Figure 3. Variation in temperature profile with time during microwave heating of a dielectric.

than surface. This phenomenon has been confirmed experimentally by many investigators working with a broad range of ceramic systems and has been discussed elsewhere [1]. The magnitude of the profile is dependent upon many factors such as power level, electric field intensity and material properties including lossiness, thermal and electrical conductivity, etc. Understanding the role played by these factors, and thus control over the temperature profile, is one of the main research directions currently being taken.

The nature of this power deposition and subsequent inverse temperature profile appear to offer a number of advantages over conventional sintering techniques, including:

3.1 The possibility of heating components very rapidly and uniformly due to a reduction in thermal stress via generation of a uniform temperature profile

Conventional sintering relies on energy reaching the surface of the ceramic component by convection (low temperatures) and radiation (high temperatures) from where it is conducted towards the cooler centre of the body. Since most ceramic materials are good thermal insulators the rate at which the energy in the form of heat penetrates a large body can be very slow leading to severe temperature gradients. As a result of thermal expansion and the brittle, unforgiving nature of ceramics this can mean that the surface, which expands more than the centre, cracks and even breaks away from the component. Heating rates are therefore severely limited, often to only 5–10°C per minute or less.

Control of the inverse temperature profile during microwave sintering so that it is virtually uniform across the ceramic allows the use of much higher heating rates since differential expansion is not experienced. There are two primary methods of achieving this. One is to use a technique referred to as 'casketing' [4] which involves surrounding the component by microwave transparent insulation such that heat losses from the surface are minimised, generating a more uniform temperature profile. The alternative is to use a combined microwave/conventional heating system so that both the interior and surface are heated. This could be achieved by designing a furnace which contains conventional heating elements in addition to a magnetron, however, the more usual approach involves the use of a microwave susceptor. The latter not only aids in heating the material at low temperatures when losses are low, but if it continues to absorb microwaves throughout the process will radiate heat on to the surface of the sample. Figure 4 shows two ferrite* specimens sintered at 1220°C for 20 minutes and employing a heating rate of approximately 60°C per minute. One was heated using primarily microwave energy but with a small element of conventional radiant heating to generate a uniform temperature profile via the use of a beryllia/silicon carbide† susceptor. The silicon carbide particulates in the latter acted as microwave absorbers whilst the high thermal conductivity of the beryllia ensured uniform heating of the susceptor. The second sample was sintered using purely conventional sintering techniques and has undergone catastrophic failure.

It should be noted that the above comments apply to heating cycles. Cooling can present greater problems than heating since now the surface of the material contracts more than the centre and is therefore placed under tension. Since brittle materials are notoriously weak in tension, this often restricts cooling rates to as little as 3–5°C per minute if thermal stresses are to be avoided. However, once again by controlling the temperature profile during the cooling cycle via the parameters mentioned above, a slightly faster cooling rate can often be tolerated. The gain is, however, generally less than that achievable during heating.

The fast heating rates attainable also reduce the possibility of obtaining grain growth. For example, zirconia has been sintered to approximately 95% of theoretical whilst retaining a submicron grain size using microwave sintering. The 3 mol% yttria doped zirconia‡ was heated to 1385°C with no hold time using a heating rate of approximately 30°C per minute. The material was fully tetragonal after cooling slowly to room temperature.

---

*Strontium hexaferrite, Philips Components Ltd., Southport, UK.
†Consolidated Beryllia Ltd., Milford Haven, UK.

Figure 4. Strontium hexaferrite after heating at 61°C min⁻¹, (a) using microwave heating and (b) using radiant heating.

3.2 The potential for sintering ceramics 'from the inside out' via generation of an inverse temperature profile

Current bulk polycrystalline ceramic superconductors suffer from the major disadvantage that only the surface layers have the correct oxygen stoichiometry and are, therefore, superconducting. This occurs because conventional sintering results in the surface densifying before the interior, cutting off the latter from the oxygen-rich atmosphere required during sintering to achieve the correct oxygen stoichiometry. This effectively creates a 'shell' of superconducting phases surrounding a central core which is non-superconducting [5]. It is this effect which has helped drive research in recent years towards the production of thin films.

Table 2. Processing conditions and properties of microwave and conventionally processed YBCO superconductors

Sintering temp/°C	Sintering time/h	% theoretical density	T_c K	Oxygen level	
				Surface	Interior
Microwave processed samples					
750	0·75	74	—	6·56	6·56
800	0·75	79	—	6·80	6·80
900	0·75	89	91	7·00	7·00
800	2	84	—	6·80	6·80
800	4	89	—	6·80	6·80
Conventionally processed samples					
950	12	81	72	6·80	6·56
970	12	85	78	6·80	6·56
970	18	89	86	6·80	6·56

Research on yttrium-barium-copper oxide (YBCO) superconductors§ has shown that the use of microwave energy resulted in a significantly more uniform oxygen content throughout disk shaped components by generating an inverse temperature profile during sintering [6] (see Table 2). The magnitude of the profile was controlled by a casketing approach in which the degree of insulation was variable (Figure 5). As a result of the difference in temperature profile, the microwave sintered samples also displayed a slight increase in grain size towards the centre of the compact, whilst the conventionally sintered samples showed the opposite behaviour.

It will be noted that sintering has occurred at lower temperatures and in shorter times when compared with conventional sintering techniques. The reason for this phenomenon, which has been reported by many investigators (*e.g.* Refs. 7–9), is still not certain, however, a number of theories have been put forward. These have been discussed elsewhere [1].

All the samples were only compacted and sintered once leading to low oxygen contents with the conventionally sintered pieces and hence low T_c values. However, the significantly more uniform oxygen content in the microwave processed samples noted earlier has resulted in values more normally associated with YBCO material. This appears to indicate that the use of microwave energy for sintering eliminates the need for grinding processes which are both time consuming and introduce impurities from abrasion of the grinding media and vessel walls.

Finally, the use of microwave energy also permitted heating rates of between 15–20°C min⁻¹ to be used, these were approximately five times faster than could be achieved in the conventional furnace if cracking was to be avoided.

Another advanced ceramic system in which it would be desirable to maintain a particular atmosphere throughout the sintering compact is in the nitridation of silicon compacts. This is currently under investigation at Nottingham and although only a handful of results have been obtained, early indications are very promising.

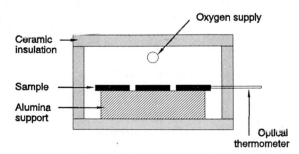

Figure 5. A schematic diagram of the casketing arrangement.

‡Toyo Soda, Tokyo, Japan, TZ3YB.
§SS-ACS YBCO powder, Superconductive Components Inc., Ohio, USA.

4. TEMPERATURE MEASUREMENT

There are many difficulties associated with temperature measurement during microwave sintering. The two principal reasons for this are; (i) the effect of microwaves on thermocouples — they induce currents, and (ii) the existence of the inverse temperature profile. This latter means that the surface temperature, which most temperature measurement devices record, is lower than the internal temperature, which controls most diffusive processes.

Thermocouples can only be used if shielded from the microwaves by a metal tube, often platinum to withstand the high temperatures generated during sintering. A metallic shield, however, disturbs the microwave field pattern in the cavity usually resulting in unwanted effects. In addition, if the tip of the thermocouple is to be in contact with the surface of the component during sintering, then to prevent electrical discharge an electrically insulating layer needs introducing between thermocouple and sample. This, however, reduces the sensitivity and accuracy of the temperature measurements.

Pyrometry is often used, however this monitors surface temperatures and, as has already been shown, these can be significantly different from internal temperatures. Nevertheless, the same is often true during heating and cooling of conventionally heated pieces and allowances can be designed into systems. It is believed that ultimately pyrometry is the most likely approach to be taken if and when microwave sintering becomes an industrial reality. However, for research purposes, where measurement of temperature profiles is required, pyrometry leaves much to be desired.

A recent alternative to this problem has been the introduction of optical thermometry in which the temperature of the sample is sensed by either a microwave transparent black body tip mounted on a sapphire rod or by using a sapphire rod as a light pipe and monitoring temperature by pyrometry [10]. (High purity sapphire has an extremely low dielectric loss and so is immune to microwave radiation). Both of these techniques have associated disadvantages, not least being the brittle nature of the sapphire rods and the high capital cost of the system. Nevertheless, comparative work carried out using shielded thermocouples, optical thermometry via the black body approach and conventional optical pyrometry indicate that the optical thermometry route appears to be significantly more accurate than thermocouples. There is also the possibility of drilling a small hole in the side of sample and inserting a second optical thermometer to monitor internal temperatures in addition to surface temperatures. Such experiments will be performed at Nottingham in the near future.

5. CONCLUSIONS

The nature of the temperature profile developed within materials heated by microwave energy has been shown to permit rapid and uniform heating and to allow ceramics to be sintered from the 'inside out' where appropriate. The former will be particularly useful when it is required to achieve a particularly fine grain size or to heat large components uniformly to avoid cracking due to thermal stresses. The ability to heat samples with a controlled inverse temperature profile is applicable when densifying components which require a gaseous phase to be in intimate contact with the material throughout the process. It can be used to generate more uniform composition throughout the structure than is achievable by conventional sintering.

ACKNOWLEDGMENTS

We wish to acknowledge the Science and Engineering Research Council and the Department of Trade and Industry for financial support which enabled much of the work discussed to be performed.

REFERENCES

1. METAXAS, A. C. & BINNER, J. G. P., in *Advanced Ceramic Processing Technology*, Ed. J. G. P. Binner, Noyes Publications, New Jersey, 285, (1990).
2. BINNER, J. G. P., *British Ceramic Proceedings*, **45**: *Fabrication Technology*, Eds. R. W. Davidge and D. P. Thompson, 97, (1990).
3. SUTTON, W. H., *Am. Ceram. Soc. Bull.*, **68**, 376, (1989).
4. HOLCOMBE, C. E. & DYKES, N. L., *J. Mat. Sci. Lett.*, **9**, 425, (1990).
5. EKIN, J. W., *Adv. Ceram. Mat.*, **2**, 586, (1987).
6. AL-DAWERY, I. A. H., MOON, J. R. & BINNER, J. G. P., *Processing to achieve uniform oxygen content in bulk YBCO superconductors*, to be published.
7. FANSLOW, G. E., in *Microwave Processing of Materials II*, Eds. W. B. Snyder Jr., W. H. Sutton, M. F. Iskander and D. L. Johnson, Materials Research Society Symposium Proceedings, **129**, 43, (1990).
8. KATZ, J. D., BLAKE, R. D. & KENKRE, V. M., in *Microwaves: Theory and Application in Materials Processing*, Eds. D. E. Clark, F. D. Gac and W. H. Sutton, *Ceramic Transactions*, **21**, 95, (1991).
9. WROE, F. C. R. & SAMUELS, J., previous paper this volume.
10. MERSHON, J., in *Microwaves: Theory and Application in Materials Processing*, Eds. D. E. Clark, F. D. Gac and W. H. Sutton, *Ceramic Transactions*, **21**, 641, (1991).

Planar Ceramic Capacitor Arrays: Aspects of Manufacturing and Processing

J. CHAMBERS and M. MARSDEN

Oxley Developments Co. Ltd., Priory Park, Ulverston, Cumbria, LA12 9QG

ABSTRACT

Modern electronic equipment is increasingly subjected to the effects of electromagnetic interference (EMI), also referred to as radio frequency interference (RFI). Unwanted electrical signals from sources such as motor commutator noise are introduced to electronic circuits by interconnecting wires, resulting often in disturbance of the circuit's normal operation. Ceramic capacitors in the form of feedthrough elements have traditionally been utilised to divert (or filter) RFI currents to ground, being mounted at the equipment interface between each conductor and the equipment case.

High relative dielectric constant (> 3000) ceramics based on barium titanate are utilised to provide a high capacitance per unit volume, albeit at the expense of temperature stability. The higher the capacitance, the lower the RF noise frequency that can be dealt with by the capacitor (usually 10's or 100's of kHz upwards). Alternatively, the higher the dielectric constant, the smaller the package that can be used to accommodate a particular capacitance. To further reduce the physical size of RFI filtering capacitors, single monolithic bodies containing multiple ceramic capacitors are utilised, known as 'planar capacitor arrays.' This paper will describe new approaches to the manufacture of planar capacitance arrays, including machining of the green ceramic body. The influence on electrical performance by the choice of electrode material will be discussed, particularly with regard to insertion loss and cross-talk performance. Further techniques for the introduction of Z direction conductive vias, perpendicular to the electrodes, will also be described.

1. INTRODUCTION

Modern electronic equipment is increasingly subject to the effects of electromagnetic interference (EMI). This is often referred to as Radio Frequency Interference (RFI) and is yet another facet of the environmental pollution created by mankind today. Many types of electronic equipment are subject to EMI and specifications have defined interference levels for military, avionic and telecommunications applications for many years.

Forthcoming legislation from the EEC (EMC Directive 89/336/EEC) will soon apply to all electronic equipment including the so called "white goods" or consumer products. This directive will specify the minimum level of interference that equipment must tolerate without malfunction (susceptibility) and also the maximum level of EM field that may be generated by the equipment itself (emissions). Designers and manufacturers of all electronic equipment now face the challenge of producing equipment that complies with the directive if they wish to continue sales in the EEC.

2. DESIGNING FOR ELECTROMAGNETIC COMPATIBILITY

Electromagnetic compatibility (EMC) of equipment has been developed to include shielding and gasketing to reduce the effects of radiated fields, and filtering techniques to remove conducted interference currents. This interference could be generated by adjacent motor commutators, fluorescent lighting and a wide range of industrial machinery. A typical method of preventing interference due to conducted currents is to use feed-through capacitors which provide a low impedance path to ground for the high frequency interference. This type of filter can easily be mounted directly in the wall of a shielded enclosure, or may be incorporated into connectors through which the input/output signals are carried to the outside world. A typical feed through capacitor will have a low self-inductance and this allows frequencies into the GHz region to be decoupled to ground without self-resonance due to L-C circuit combinations.

Feed-through capacitors can be of the tubular (coaxial) or multi-layer discoidal type. Figure 1 shows the cross-section of these two devices. The tubular type is ideally suited for making Pi-section filters by simply incorporating an inductive element within the tube, but the capacitance per unit volume is limited with this type of structure.

The discoidal capacitor on the other hand can have many layers of interleaved electrodes and this leads to high capacitance per unit volume. The capacitance of these devices can be greatly increased by the use of dielectric mediums with a high relative permittivity (E_r). This can be achieved by using ceramics based on barium titanate in which E_r can be $> 10,000$. A major problem with these materials is that the temperature co-efficient of capacitance is very high. As a result the X7R series of dielectrics has been developed which has an E_r value of ≈ 4000 and an improved capacitance variation of $\pm 15\%$ over the temperature range $-55°C$ to $125°C$.

2.1 Filter performance

The performance of a filter system is commonly referred to as Insertion Loss (I.L.). This is the ratio of the voltages across a load with and without the filter in the circuit and is normally expressed as a logarithmic ratio in decibels (dB). For each capacitive or inductive element within a filter, the resultant theoretical I.L. is increased by 20 dB per decade. A simple L-C filter will give an I.L. of 40 dB per decade and a pi section (C-L-C or L-C-L) will give 60 dB per decade. These results are modified in the real world due to parasitic

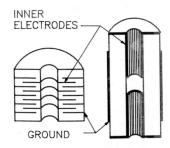

Figure 1. **Cross section of discoidal capacitor (left) and tubular capacitor (right) showing electrode configuration.**

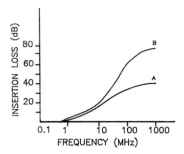

Figure 2. Typical insertion loss curves for (a) 5000 pF capacitor filter, (b) 5000 pF pi-section filter.

inductance, capacitance and resistance, resulting in I.L. "levelling off" at approximately 60–80 dB. This is illustrated in Figure 2 showing typical insertion loss/frequency for 5000 pF capacitor and pi section filters.

2.2 Limitations of existing technology

A major limitation of this type of filtering technology is the amount of space taken up by these discrete components and the wiring interconnections required. To overcome this limitation, filters have been incorporated into the connectors used to interface the equipment to the outside world. This is partly to reduce interconnections but the main advantage is in reduction of space requirements and total weight, particularly important in avionic applications. A further limitation is then imposed by the pitch between adjacent pins in the connectors. This limitation has now been further reduced by producing single monolithic bodies containing multiple capacitor arrays, known as "planar capacitor arrays."

3. PLANAR CAPACITOR ARRAYS

The basic structure of a planar array is very similar to the multi-layer discoidal and is illustrated in Figure 3. In this, multiple layers of ceramic dielectric are built up with overlapping electrodes in between. The electrodes are screen printed using a palladium or palladium/silver ink depending on the requirements. The multi-layer ceramic is then formed to its final shape in the 'green' stage before sintering. Sintering takes place at temperatures up to 1400°C when the electrodes and ceramic form a single co-fired monolithic structure. The finished array would typically be manufactured so that the capacitor positions coincide with the contact positions of a multi-way connector. The common ground electrodes would extend to the periphery of the array where a ground connection would be made via the connector body. The whole assembly can then capacitively decouple each line individually with a greatly reduced assembly time compared with discrete components.

The planar array is also suitable for filtering connectors in which lines have different filtering requirements. The designer can arrange for certain lines to have different capacitance values by altering the electrode patterns, to ground

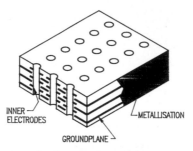

INNER ELECTRODES
GROUNDPLANE
METALLISATION

Figure 3. Cross section of planar array.

some lines by extending the common ground planes to the through pins or provide no filtering by removing all electrodes from a particular pin. This is then a "custom designed" array and limitations in the timescale required to produce such an array may be a problem.

Machining or forming of green ceramic has been traditionally carried out using punch tools. The need to manufacture these tools produces a long lead time before components can be produced. This limitation can be overcome by using modern, flexible manufacturing methods utilising CNC machines. The significance of this becomes apparent when one considers that the pressures used on punch tools can be of the order of 20–100 MPa and consequently these tools must be made to withstand these forces and maintain tight tolerances on the dimensions of the finished article.

There are additional limitations introduced by the use of planar arrays in terms of both mechanical and electrical performance. The dielectric ceramic (barium titanate) is a brittle material with poor thermal conductivity. Considerable care must be taken when soldering to the planar array to avoid excessive thermal gradients and subsequent mechanical damage such as cracking. Both controlled heating and mechanical connection methods can be utilised to overcome these problems.

4. MULTI-LAYER CAPACITOR PROCESSING

The processing route for manufacturing ceramic MLCs begins by blending the ceramic powder with an organic binder. After milling to ensure a homogeneous slurry, tape is cast in a continuous film. The tape is then cut into manageable lengths, carefully graded and passed on for screen printing. To ensure accurate, repeatable screen printing, this can be controlled using computer aided video displays to compare the printed electrode pattern with the required image. The conventional route would then be to laminate and punch out the components using a press tool prior to burn-off of the binder, followed by final sintering of the components at 1400°C.

4.1 Flexible manufacturing technology

The use of CNC machining facilities to drill and shape green ceramic tape has considerably reduced the lead time to produce components from a new design. Initial lead time to produce prototypes is reduced and the opportunity to make

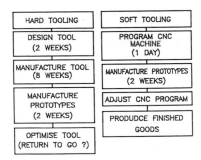

Figure 4. Hard tooling vs. soft tooling.

dimensional changes without major tooling changes is retained. A typical conventional manufacturing route for a planar array is shown in Figure 4 and contrasted with the route for a soft tooled (CNC) manufactured component.

5. LIMITATIONS OF PERFORMANCE

The electrical performance of a planar array, when compared with individual capacitors, may be compromised due to the difference in grounding paths. The performance of a filter at high frequencies can be seriously impaired by the introduction of a small resistance to ground. This occurs as the impedance of the capacitor approaches zero with increasing frequency and the ground resistance then becomes significant. The effect is to shunt the noise through the filter rather than passing it down to ground. If an individual capacitor is mounted directly to a metal bulkhead of effectively zero impedance, its ground path is short and ground currents from adjacent capacitors will not cause local ground plane potentials. In the case of a planar array, the local ground plane consists of a thick film electrode $\approx$ 10 microns thick, and consequently ground plane potentials may exist due to ground currents from adjacent filter lines. The effect of this is an apparent reduction of filter performance. Steps can be taken both in the design and manufacture of planar arrays to reduce this effect. These include the use of high conductivity ground plane materials and careful attention to the outer termination of the ground plane. For critical applications, novel methods are possible to further reduce ground plane resistance.

6. NOVEL APPLICATIONS

One of the notable features of the flexible manufacturing technology described previously is the ability to insert vias in the Z-plane during the green ceramic processing. Figure 5 illustrates the technique where vias placed between the ground plane electrodes bring low inductance connections to the surface. The top surface can then be metallised after sintering and electroplated with copper and silver to provide a low resistance external ground plane. This technique allows improved filter performance to be achieved in demanding applications.

A further use of Z-plane vias is shown in Figure 6 where an array of non-

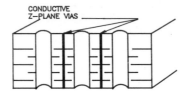

Figure 5. Internal ground planes connected to top and bottom surfaces in Z-plane using conductive ink.

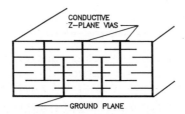

Figure 6. A capacitor array utilising Z-plane vias with a common ground plane also using Z-plane vias.

feedthrough multilayer capacitors is provided with a common ground electrode. In this case the vias are used to interconnect the internal electrodes and bring these connections to the surface. A device of this type would typically be mounted directly on a printed circuit board where the dimensions of the array would be tailored to suit the board.

7. CONCLUSIONS

Planar capacitor arrays offer a means of combining multiple capacitor locations within one ceramic body. Care must be exercised in their specification and assembly in order to obtain the maximum benefit from their use. The use of CNC machines in their manufacture eliminates the long lead times associated with conventional press tooling, and allows design optimisation without seriously extending lead times. Insertion of Z-planes to reduce ground paths can give improved filtering performance.

Advances in PTC Thermometrics Technology

J. H. McCARTNEY, R. E. W. CASSELTON, C. M. MORTER and
J. C. HOLLOWAY
Bowthorpe Thermistors, Taunton, Somerset, TA2 8QY, U.K.

ABSTRACT

An important application of PTC (positive temperature coefficient of resistance) thermistors is in the protection of telephone exchange equipment against high voltages and currents induced into the telephone lines by various fault conditions. The thermistor works by heating up rapidly when a heavy current passes through it and limiting the current to a low value as it goes into its high resistance state. When the fault is cleared the thermistor cools down and returns to its original state, i.e. it resets and does not need to be replaced. The thermistor disc does not heat up uniformly so stresses are generated which, if not controlled, can cause fractures. The challenge is to improve both the electrical and the mechanical properties of the ceramic as well as optimising the design of the thermistor itself. This is being achieved by a combination of theoretical studies and practical work. The aim is to understand the nature of the stresses, refine the ceramic grain structure, reduce the physical inhomogeneities in the pressed discs, and explore novel design configurations.

1. INTRODUCTION

PTC (positive temperature coefficient) thermistors are electronic components based on ceramic barium titanate. The material, which is doped to make it conductive at normal temperatures, exhibits a steep increase in resistance of several orders of magnitude near the Curie Temperature. An important application is in the protection of electrical circuits against excessive voltages and currents. When such faults occur the thermistor is subjected to both electrical and mechanical stresses. This paper describes how the thermal shock resistance has been increased by improving the ceramic and by exploring new design configurations.

For reviews of the theory and origin of the positive temperature coefficient of resistance effect in barium titanate, and further details of the design principles of PTC thermistors, see references [1] and [2].

2. PTC THERMISTOR APPLICATIONS

One of the first applications was in temperature sensing. The sharp increase in resistance (Figure 1) allows the attainment of a predetermined temperature to be detected. This temperature is around 120°C for barium titanate but can be shifted up or down by substituting lead or strontium respectively for some of the barium.

PTC thermistors are also used in colour television degaussing circuits. When mains voltage is applied to a thermistor in series with a coil placed around the tube a heavy current passes through the coil initially producing an alternating magnetic field that decays as the thermistor heats up.

They are used as heating elements in a wide range of domestic appliance and automobile applications. As the ambient temperature changes, the current

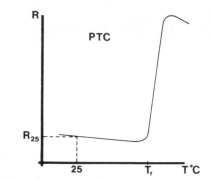

Figure 1. Resistance-temperature characteristic.

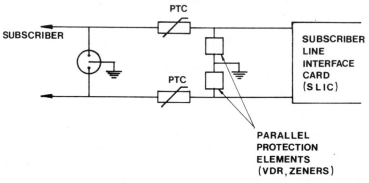

Figure 2. Telephone exchange protection circuit.

through the element changes in such a way as to compensate, so such elements are self-regulating. Heaters account for by far the greatest consumption of PTC ceramic material.

Bowthorpe Thermometrics specialises in developing and designing devices for use in telecommunications equipment. PTC thermistors are used to protect line cards, battery feed and ringer circuits at the exchange end of a telephone line against heavy currents and high voltages entering the telephone line by induction or by direct contact with a power cable. They also protect handsets, modems and similar equipment at the subscriber end.

Figure 2 is a schematic diagram of a telephone exchange protection circuit. The thermistor is used with other components to provide protection to electronic circuits in the exchange against a range of electrical hazards. The gas discharge tube, for example, conducts high voltage pulses from lightning strikes to ground. The thermistors limit heavy currents and hold off high voltages caused, for example, by a mains cable falling across a telephone line. The parallel elements ensure that the residual fault voltage seen by the protected circuit never exceeds a safe limiting value. They also react very quickly, conducting the fault current to ground for the brief period that the thermistors take to heat up.

Figure 3. PTC thermistors.

3. THERMISTOR DESIGN

PTC thermistors are typically in the form of ceramic discs provided with electrodes to which wires are soldered (Figure 3). In some applications the disc may be held between spring contacts or soldered directly to a substrate.

In designing a protection thermistor it is necessary to ensure that it will "trip" whenever a minimum fault current passes through it, but that it will not trip accidentally with normal operating currents. The thermistor must be able to withstand the mechanical stresses resulting from the high heating rates generated by the fault current and voltage. Finally it must withstand the maximum fault voltage for an extended period of time without breaking down.

Figure 4 shows the equilibrium current-voltage characteristic for a PTC thermistor. It is obtained by slowly increasing the voltage across the device and measuring the current flowing through it. At any point on the curve the joule heating power ($I \times V$) is balanced by the power dissipated as heat into the surrounding air and down the wires. The current maximum is an important parameter because it determines the non-tripping/tripping characteristics of the thermistor. Its value is influenced by factors such as the resistivity of the ceramic and its switch temperature, the surface area of the disc and the length, diameter and material of the lead-wires or other form of electrical contact. All these factors are used as parameters in the design of a PTC thermistor.

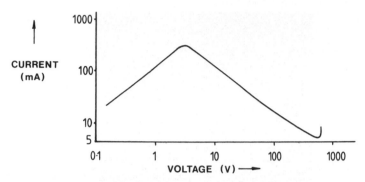

Figure 4. Equilibrium current-voltage characteristic.

Further along the I-V curve the current drops to a minimum, corresponding to the maximum in the resistance-temperature characteristic. As the voltage is increased beyond this point thermal runaway occurs and the device is destroyed. In designing a thermistor a suitable margin is allowed between the maximum fault voltage and the runaway voltage.

Early PTC thermistors were very limited as to the levels of current and voltage they could handle without breaking down. A particular problem was that the resistance-temperature curve was extremely voltage sensitive. The resistance increased by some four orders of magnitude above the Curie Temperature when measured with a small electric field, but by only two orders of magnitude when a typical fault voltage was applied.

Above the Curie Temperature only the grain boundary regions have a high resistivity; the resistivity of the core of each grain remains low. So a voltage applied across the thermistor is dropped stepwise across the grain boundaries. The potential barrier at each grain boundary, which contributes to the origin of the PTC effect, is reduced by an amount proportional to the voltage across it, and the total resistance is reduced accordingly.

Matters were greatly improved by refining the ceramic grain structure. The reduction in average grain size increased the number of grain boundaries so that the voltage drop across each was lower. Figures 5(a) and 5(b) show typical structures for early and currently manufactured ceramic discs. The zero-field resistance range (ratio of maximum to minimum resistance) increased from four to six orders of magnitude, and the high-field range from two to five orders. The maximum sustainable electric field increased fourfold. This improvement meant that a thermistor of a given rating could be made thinner, resulting not only in a saving in materials cost but also a reduction in switching response time. In addition the finer grained ceramic was much stronger.

The much enhanced performance was achieved by careful attention to the chemical and physical properties of the starting materials, and a systematic optimisation of dopant concentrations and manufacturing process conditions. To maintain a consistent product quality from what became a highly-tuned sequence of process stages it was found essential to identify all the significant

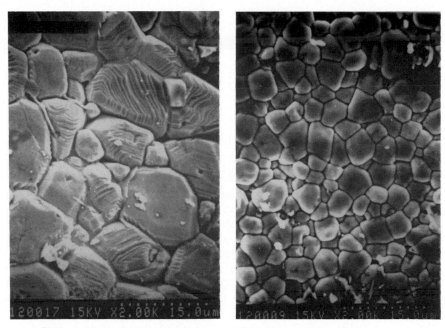

Figure 5. Grain structures of (a) early and (b) current PTC ceramics.

parameters that could possibly affect the properties of the ceramic and then to measure them regularly, or continuously where appropriate. Only in this way did it become possible to spot process deviations and apply corrections in a timely manner.

4. TEMPERATURE AND ELECTRIC POTENTIAL DISTRIBUTIONS IN A THERMISTOR DISC

It was useful to analyse the distributions of temperature and potential that develop within a thermistor disc as it heats up when a current is passed through it. This was done by creating a finite element model of the thermistor, taking into account the electrical and thermal properties of the ceramic.

Figures 6(a) and 6(b) respectively show lines of constant potential and temperature in a section through the centre of the disc, perpendicular to the electrodes, at the point in time at which it is starting to switch into its high resistance state. A heavy current has heated the disc rapidly but the core has reached a higher temperature than the rest, causing it to switch into a high resistance state. Consequently almost all the applied voltage is dropped across this core region. The core region extends almost to the cylindrical surface, creating a severe temperature gradient just under the surface. It is at this point in the heating cycle that the mechanical stresses are greatest and it has been confirmed from electrical measurements that if failure is going to occur it will happen at this instant. Eventually heat is conducted away from the core to equalise the temperature throughout the disc. The rest of the material becomes high in resistance so that heat is generated more uniformly.

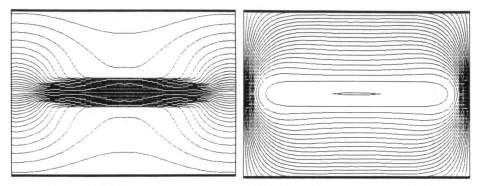

Figure 6. (a) potential and (b) temperature distribution near the Curie Temperature.

When failure occurs in practice it does indeed often take the form of a crack parallel to the electrodes near the mid-section of the disc. This is especially true where the thermistor has to be designed to use a thick disc to withstand very high voltages. Experimental work has confirmed that the crack propagates from the surface. It has also been found that the likelihood of cracking is influenced by various chemical and physical treatments to the cylindrical surface, which supports the idea of the cracks being initiated by surface defects.

In a subsequent upgrading of the model, account has been taken of the influence of the solder and lead-wires. This has been useful in the failure analysis of thin discs which tend to fracture near the solder joint.

5. IMPROVING PERFORMANCE

In principle there are two approaches to increasing the ability of a thermistor to withstand heavy surge currents in combination with high transient electric fields. The first is to attempt to make the ceramic stronger, which will normally entail making changes to the manufacturing process. The second is to modify the configuration of the thermistor so that a given current/voltage combination generates lower stresses in the ceramic.

Figure 7 shows fracture failure rates for thermistors made from powder manufactured by a standard and an improved process. For each test voltage a resistor was inserted in series with the thermistor to limit the initial current to a common level.

Thermistors made from the standard powder failed at 260 V upwards. Interestingly, there is no threshold voltage below which all thermistors passed and above which they all failed: although some failed at 260 V others survived 420 V. This observation indicates a wide variation in the mechanical strength of the discs which reflects the statistical distribution of defects within the ceramic and/or at the surface.

An overall increase in failure voltage was achieved by paying careful attention to attaining the following:

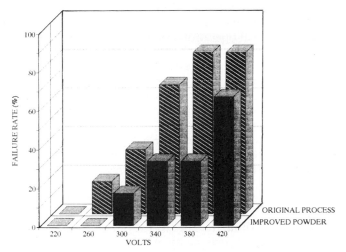

Figure 7. Fracture failure rates.

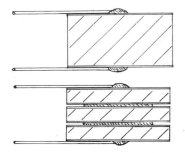

Figure 8. Single and triple disc constructions.

— a homogeneous distribution of all the components in the PTC mix formulation;

— optimal spray dried granule size distribution to ensure good die filling on pressing;

— correct press action to minimise pressing stresses in the green compact.

While not dramatic, the improvement observed in Figure 7 was sufficient to ensure an adequate margin of safety in meeting the customer's specification.

However, the wide variation in breakdown voltage remained. To improve matters here it would be necessary to review all stages of the manufacturing process, with the aim of eliminating crack-inducing defects. The potential benefit of being able to make all the thermistors perform as well as the best ones in this population is enormous. In the meantime the specification of a thermistor has to be based on the lowest level of performance.

For another application a thermistor had to be designed to protect against a particularly difficult set of current/voltage fault conditions. At the same time there were severe space limitations on the circuit board in which it was to be used. This meant that the diameter of the disc had to be smaller than the ideal. In the event, all the normal operational requirements were met but the devices performed very badly on the fault tests. Adjustments to the manufacturing process were of no benefit in this case.

The problem was solved by replacing the single disc with three thinner ones soldered together, to give a structure with the same overall diameter and thickness (Figure 8). The triple disc construction passed the fault tests easily whereas the original design failed totally.

The reason seems to be that the centre disc, where most of the heat is generated initially, is not rigidly clamped to the outer discs, particularly if the solder layers between them are confined to a small area. This allows a degree of relative expansion which is not available to the central portion of the ceramic in the single disc construction. Another factor may be that thin discs tend to be more structurally homogeneous than thick ones. Also the presence of additional electrodes within the bulk of the device, at the interfaces between the individual disc elements, may contribute to a more uniform current flow.

6. SUMMARY

Considerable enhancements in the performance of PTC thermistors used in circuit protection applications have been brought about by refining the ceramic grain structure and improving powder processing techniques. Novel device structures promise further useful advances. Mathematical modelling of the thermistor has given a valuable insight into its mode of operation and the origin of mechanical stresses.

Further fundamental work should concentrate on identifying the causes of the wide disc to disc variations in mechanical strength, with the aim of developing cost-effective ways of manufacturing a consistently stronger ceramic.

ACKNOWLEDGMENTS

The modelling work was done by the Mathematical Institute, Oxford University.

The authors wish to thank the Directors of Bowthorpe plc. for permission to present and publish this paper.

REFERENCES

1. NOWOTNY, J. & REKARS, M., *Ceram. Int.,* **17**, 227, (1991).
2. SINCLAIR, D. C. & WEST, A. R., *J. Mat. Sci. Lett.,* **7**, 823, (1988).

The Mechanical Behaviour of Multilayer Ceramic Capacitors — Applied Stresses and Testing Techniques

R. FREER,* R. AL-SAFFAR,* I. TRIBICK† and C. P. WARD†

*Materials Science Centre, University of Manchester/UMIST, Grosvenor Street, Manchester, M1 7HS

†AVX Ltd., Hillman's Way, Coleraine, Northern Ireland, B52 2DA

ABSTRACT

Multilayer ceramic (MLC) capacitors are subject to a variety of stresses during manufacture, during their mounting on electronic circuit boards and in use. A brief review is given of the origins and types of stress, and the tests used to determine mechanical, thermal and elastic parameters of MLC capacitors. Representative data are presented and typical trends identified.

1. INTRODUCTION

The continued drive to reduce the size of discrete components, such as multilayer ceramic (MLC) capacitors, used in electronic systems has placed increasing demands upon the performance and reliability of the devices. Whilst ceramic processing techniques are allowing the manufacture of thinner dielectric layers and therefore a higher capacitance per unit volume, there is growing interest in the mechanical reliability of MLC capacitors, particularly in the stressful environments encountered during manufacture and circuit assembly.

In principle, MLC capacitors may be regarded as monolithic 'units' for the purpose of mechanical testing, but strictly they contain at least three components: ceramic dielectric (frequently based on $BaTiO_3$), internal electrodes (often Pd or Pd-Ag alloy, depending upon the ceramic sintering temperature) and termination electrodes (which may involve several layers, ending with Pb-Sn solder). The intrinsic mechanical properties of MLC capacitors depends upon the type of materials employed, their relative proportions, the design geometry and processing conditions. Individual multilayer capacitors should have structural integrity with no interconnecting porosity, cracks, voids, and no delamination of the respective metal-ceramic layers [1, 2].

2. STRESSES APPLIED TO MLC CAPACITORS

During manufacture MLC capacitors are subjected to a varity of stresses. The sintering of the multilayer metal-ceramic compacts results in mechanical stresses from thermal expansion mismatch between the metal and ceramic components, and this may be enhanced by the presence of impurities having different expansion characteristics to the host phase [3]. The evolution of gases during sintering, structural phase changes of the dielectric and grain size effects may also generate stresses [4, 5, 6].

Tumbling of MLC capacitors is performed to round the corners after sintering, but the mechanical action may occasionally cause minor surface damage. More severe are the thermal stresses developed during the application of terminations, and during wave soldering. Thermal shock behaviour is a complicated problem and has been considered by several authors [7-10]. More general treatments of mechanical properties are given by Frieman and Pohanka [3] and Koripella [10].

Once the MLC capacitors are in use there are potential mechanical stresses from flexure of circuit boards, or indeed fracture fatigue in high vibration environments. Thermal cycling or large changes in temperature can cause changes in the intrinsic mechanical strength, sometimes because of structural phase changes [5, 3, 11]. For small components the local electric fields inside MLC capacitors are usually modest, but high field levels may be achieved in power electronic systems or at times of power surges. Whilst high fields may cause local, though not necessarily destructive, dielectric breakdown, there is evidence that the presence of an electric field can enhance the mechanical strength of capacitors based upon piezoelectric barium titanate [12].

Other important environmental factors include relative humidity since capacitor ceramics tend to undergo slow crack growth in humid environments [13].

3. TEST PARAMETERS AND TEST TECHNIQUES

3.1 Modulus of Rupture

Modulus of rupture (MOR) or flexure strength tests have been used extensively for the evaluation of MLC capacitors [10, 12, 14-17]. The simplicity of the three point loading technique means that it can easily be employed for large [12] or small [13-16] components. Four point loading geometry was used by Vora and McHenry [18], but the technique becomes increasingly more difficult to use for the smaller size of devices, *e.g.* 1206. Since the flexure strength is not constant for a particular material, depending upon the probability that a flaw capable of initiating failure will be present at specific stress, it is convenient to analyse the results in terms of Weibull distribution functions [3, 10, 19].

3.2 Fracture Toughness

The most convenient method to determine fracture toughness for small chip capacitors is via indentation-fracture techniques [20] where a flaw of predictable size is created by a simple Vickers hardness indentation. The fracture toughness K_{IC} is then obtained from the measurement of flexure strength (σ_f) in a material containing an indent generated by an indentation load (P)

$$K_{IC} = \eta(E/H)^{1/8}(\sigma_f P^{1/3})^{3/4} \tag{1}$$

where E is Young's modulus, H is hardness and η is an empirical constant (~ 0.59) [20]. Strictly, Young's modulus is required for the multilayer capacitor environment, rather than the dielectric alone. Fracture toughness data for different MLC capacitors have been obtained by several workers, *e.g.* [10, 11, 14, 16, 18].

3.3 Thermal Shock

The weakening of ceramic components by a rapid change of temperature is of immense practical importance for MLC capacitors, but there is as yet no commonly agreed method of assessing the effect of thermal shock. Koripella and De Matos [21] demonstrated the damage that can result from the thermal shock of capacitors, and Power [22] described a factorial analysis experiment designed to minimise the damage caused by wave soldering. Koripella [10] utilised thermal shock resistance parameters R and R^1 to evaluate the behaviour of MLC capacitors:

$$R = \Delta T = \sigma_f(1 - v)/E\alpha \qquad (2)$$

and

$$R^1 = \sigma_f(1 - v)k/E\alpha \qquad (3)$$

where α is the coefficient of thermal expansion, k is the thermal conductivity, v is Poisson's ratio and other symbols (E and σ_f) have the meaning defined previously. High values of R^1, denoting good thermal shock resistance (to avoid fracture initiation), are associated with high fracture stress (σ_f) and thermal conductivity, and low values of Young's modulus and coefficient of thermal expansion.

Koripella [10] also showed that further understanding of such problems can be obtained by thermally shocking MLC capacitors at different temperatures and then performing standard MOR tests. The critical ΔT (difference in temperature) above which the mechanical properties are severely degraded, may be readily defined by such techniques.

3.4 Other Mechanical Tests

A variety of other mechanical testing procedures have been employed including double torsion and applied moment double cantilever beam fracture toughness [16], compression, and edge flaking tests [15].

3.5 Elastic Properties

For the evaluation of fracture toughness (K_{IC}) and thermal shock resistance (R^1), selected elastic moduli are required, namely Young's modulus for K_{IC} and Young's modulus plus Poisson's ratio for R^1. Whilst relevant data may be obtained from strain gauges attached to samples undergoing MOR tests [14], an alternative procedure is to use ultrasonic techniques to determine longitudinal and shear wave velocities and from these calculate elastic parameters [23].

4. EXPERIMENTAL DATA

In Tables 1–5 representative data for mechanical, elastic moduli and thermal properties of ceramic dielectrics and MLC capacitors are listed. The data are taken from different sources and relate to different components and compositions. Comparison should not be made between the tables, but the

**Table 1. Flexure
strength of (1206)
dielectric blanks [10]**

Dielectric	MOR (MPa)
COG	281
X7R	209
Z5U	164

**Table 2. Flexure strength of (3220)
dielectric blanks and multilayer ceramic
capacitors [24]**

Dielectric	MOR (MPa)
Z5U blank	102
X7R blank	130
As fired X7R MLC with 17 electrodes	145
Terminated X7R MLC with 17 electrodes	180

**Table 3. Fracture toughness of MLC
capacitors [3]**

Designation	Manufacturer	K_{IC} (MPa.m$^{\frac{1}{2}}$)
NPO	AVX	1·4
X7R	AVX	0·8–1·1
Z5U	AVX	0·73–0·90
T 3000	Du Pont	1·0
BL172 HD	Du Pont	1·5

data within each table illustrate general trends. For example, flexure strength and fracture toughness for dielectric "blanks" (*i.e.* ceramic without internal electrodes) and MLC capacitors are generally in the sequence COG > X7R > Z5U. Young's modulus data follow the same sequence, but Poisson's ratio shows little variation with composition. However, thermal conductivity values are in the reverse order with Z5U > X7R > COG. Further data are available in the references cited at the end of this paper.

The experimental data alone enable comparisons to be made between different materials, and the effect of specific processing steps or variables, but considerable insight into the behaviour of the materials may be gained from detailed fractography studies after mechanical or thermal tests, *e.g.* [12, 21, 24].

Table 4. Elastic properties of dielectric blanks and MLC capacitors

Dielectric	Young's modulus E (MPa × 10⁵)	Poisson's ratio ν	Ref.
COG blank	1·273	0·35	10
X7R blank	1·067	0·35	10
Z5U blank	0·853	0·35	10
NPO-1 MLC	2·06	0·28	23
NPO-2 MLC	2·37	0·33	23
BX MLC	1·43	0·37	23

Table 5. Thermal properties of dielectric blanks [10]

Designation	Thermal diffusivity (m² s⁻¹ × 10⁻⁸)	Specific heat (Cp) (W s/g K) at 300°C	Thermal conductivity (W/m K) at 300°C
COG	ʹ5·9	0·5226	1·29
X7R	59·2	0·4803	1·66
Z5U	63·8	0·5134	1·79

5. CONCLUSIONS AND PROSPECTS

This brief survey has reviewed testing techniques which are relevant to an understanding of the mechanical behaviour of MLC capacitors. Simple trends in the mechanical properties can be identified as a function of composition/ dielectric type.

For the future there will be the challenge to determine the mechanical properties of the smaller component sizes, *e.g.* less than 1206. It will be interesting to see if mechanical integrity and strength can be maintained as the dielectric layers become thinner, and also as new dielectric formulations and processing techniques become available.

REFERENCES

1. WARD, C. P., *Br. Ceram. Proc.,* **41**, 85, (1989).
2. PEPIN, J. G., BORLAND, W., CALLAHAN, P. O. & YOUNG, R. J. S., *J. Amer. Ceram. Soc.,* **72**, 2287, (1989).
3. FRIEMAN, S. W. & POHANKA, R. C., *J. Amer. Ceram. Soc.,* **72**, 2258, (1989).
4. POHANKA, R. C., *Ferroelectrics,* **10**, 231, (1976).
5. POHANKA, R. C., FRIEMAN, S. W. & BENDER, B. A., *J. Amer. Ceram. Soc.,* **61**, 71, (1978).
6. RICE, R. W. & POHANKA, R. C., *J. Amer. Ceram. Soc.,* **62**, 559, (1979).
7. MARTEN, G. & DANFORD, A., *Surface Mount Technology,* **3**, 11, (1989).
8. MANN, L. A., GUPTA, S. P. & JONES, L. G., *Proc. 41st Electronic Components and Technology Conference,* Atlanta, May 11–16, 457, (1991).

 9. De MATOS, H. V. & SNYDER, W. B., *SMART III, Annual Proceedings,* SMT III-11, (1987).
10. KORIPELLA, C. R., *Proc. 41st Electronic Components and Technology Conference,* Atlanta, May 11–14, 457, (1991).
11. COOK, R. F., FRIEMAN, S. W., LAWN, B. R. & POHANKA, R. C., *Ferroelectrics,* **50,** 267, (1983).
12. AL-SAFFAR, R., FREER, R., TRIBICK, I. S. & WARD, C. P., *Proc. CARTS-EUROPE '91,* 99, (1991).
13. FRIEMAN, S. W., in *Reliability of Multilayer Ceramic Capacitors, Report NMAB-400,* National Materials Advisory Board, National Academy of Sciences, Washington DC, 205, (1983).
14. EWELL, G. J., *Proc. CARTS-EUROPE '88,* 60, (1988).
15. GEE, M. G. & MORRELL, R., *Proc. EUROMAT '91,* Cambridge, August, 1991, 13, (1991).
16. McKINNEY, K. R., RICE, R. W. & WU, C., *J. Amer. Ceram. Soc.,* **69,** C228, (1986).
17. McHENRY, K. D. & KOEPKE, B. G., *MRS Symposia Proceedings,* **72,** *Electric Packaging Materials Science II,* (1986).
18. VORA, H. & McHENRY, K. D., *Crack propagation in layered ceramic dielectrics,* Final Report Under Hughes Aircraft Company Purchase Order, 04-4116-0-SC6, May 15, (1980), Honeywell Inc. Corporate Materials Science Centre, Bloomington, Minnesota.
19. WEAVER, G., *J. Materials Education,* **5,** 677, (1983).
20. CHANTIKUL, P., ANSTIS, G. R., LAWN, B. R. & MARSHALL, D. B., *J. Amer. Ceram. Soc.,* **64,** 539, (1981).
21. KORIPELLA, C. R. & De MATOS, H. V., *J. Amer. Ceram. Soc.,* **72,** 2241, (1989).
22. POWER, O. K., *J. Amer. Ceram. Soc.,* **72,** 2264, (1989).
23. ONO, K., *Ultrasonic testing of elastic moduli of ceramics,* Final Report on Hughes Aircraft Company P.O. H4-491633-SC6, April, (1979), and P.O. S4-440568-SLI, March, (1980).
24. AL-SAFFAR, R., FREER, R., TRIBICK, I. S. & WARD, C. P., *Proc. CARTS-EUROPE '90,* 191, (1990).

Fabrication of Composite Ceramics Using a Dry Mixing Process

MARIE C. WILLSON, H. J. EDREES* and A. KERR
Structures and Materials Engineering Centre, National Engineering Laboratory, East Kilbride, Glasgow, G75 0QU
*Department of Metallurgy & Engineering Materials, University of Strathclyde, Glasgow, G1 1XN

ABSTRACT

The development of ceramic matrix composites (CMCs) has made it possible for engineering ceramics to be used for low stress, load bearing applications. A considerable amount of effort has been expended on the development of these materials, but until recently, less emphasis has been placed on the economic viability of the fabrication processes. The primary objective of this project was to study a potential route for the production of randomly orientated short fibre CMCs, via a cost effective, near net shaping process. A powder metallurgy fabrication route was identified as the best method. The composite system studied was a silicon nitride matrix with a secondary reinforcing phase of silicon carbide whiskers. The route involved: (1) A dry mixing stage which successfully produced a homogeneous distribution of powders and whiskers; (2) The production of soft agglomerates, to reduce problems related with fine powder handling; (3) Consolidation by uniaxial and isostatic pressing; (4) Densification by overpressure sintering.

Comparative strength analyses were carried out on the composite specimens using a biaxial testing method. The results indicated that this processing route, with further development, could be considered as a potential route for the production of randomly orientated, short fibre CMCs.

1. INTRODUCTION

The considerable interest in engineering ceramics in recent years has been because of their unique combination of engineering properties; namely:

— High temperature strength.
— Low density.
— High Young's modulus.
— Low coefficients of thermal expansion and conductivity.
— Corrosion and oxidation resistance.
— Wear and erosion resistance.
— High hardness.

Unfortunately coupled with these advantages are disadvantages which have resulted in a relatively limited range of applications for this group of materials, *i.e.:*

— Inherent brittleness.
— Lack of reliability.
— High fabrication temperatures.
— Diamond machining required.

To enable the UK to compete in this growing market, these property disadvantages have to be addressed both in terms of material properties and economic viability. The development of ceramic matrix composites (CMCs) is a readily identified and implemented mechanism for the production of tougher and more reliable ceramics. Much work has been undertaken on the development of properties of CMCs, with the recognition that further process development work needs to be carried out. The processing conditions are of extreme importance as the introduction of any defects in the form of inclusions, cracks or pores will lead to early failure of the final component (the defects act as stress concentrators). There are a variety of techniques for producing CMCs, the route obviously being dependent on the matrix material and the type of reinforcing phase, *i.e.* continuous fibre, particle, platelet, short fibre or whisker. The composites which this paper deals with consist of randomly orientated silicon carbide whisker reinforcement within a silicon nitride matrix. Isotropic property requirements in the sintered material make this composite relatively difficult to produce. Homogeneous mixing of the whiskers or short fibres is generally achieved using a wet blending technique, where pH levels need to be carefully controlled for each given system. This introduces an additional drying stage to the process, thus having a detrimental effect on component cost and the potential for the introduction of defects. The problems associated with the manufacture of the silicon carbide whiskers generally resulted in the balling or clumping of the material. One problem is that the whiskers are manufactured with a high aspect ratio, this is exacerbated by the growth mode which produces whisker bundles which sprout from a single source tying many whiskers together. The whiskers must therefore be broken down, by milling, in order to disentangle and individualise them. If whiskers are used with aspect ratios in excess of 25–40, anisotropic properties and excessive voidage, from poor packing densities, will result in the final material, Milewski [1]. In addition to mixing problems, consolidation techniques have to be considered, as do sintering conditions. Techniques which have been found to be successful involve wet blending, followed by expensive hot pressing or hot isostatic pressing routes. This paper outlines a potential PM technique for the production of a silicon nitride matrix with a silicon carbide whisker reinforcing phase. The technique addresses both the issues of property improvements and economic considerations. In addition to ease of process this net shape route minimises the need for expensive diamond machining.

2. EXPERIMENTAL DETAILS

The production process based on a powder metallurgy PM route and involves slight modification of the general mix/compaction/sinter route.

2.1 Process Variables

The process variables identified as being important are:

Materials Selection

— Matrix material.

— Type of reinforcing phase.
— Volume fraction of reinforcing phase.
— Aspect ratio of the reinforcing phase.
— Packing arrangement and alignment.
— Chemical compatibility (matrix and reinforcement).
— Sintering additions.
— Binder additions.
— Chemical purity of fibres/whiskers.

Pre-consolidation:
— Mixing techniques to ensure homogeneity.
— Free flowing powder properties.
— Particle size and morphology.

Consolidation:
— Consolidation techniques (unidirectional or isostatic).
— Pressing pressure (compaction ratios).

Densification:
— Sintering mechanism.
— Sintering temperature, pressure, time and atmosphere.

2.2 Production process:

Rotational turbular mixing

The silicon nitride powder (Cookson 1002) was mixed in a turbular mixer with the silicon carbide whiskers (American Matric Inc.) and the sintering additive magnesia (10 wt%). An organic binder (liquid grade PEG) was added at this stage to give the green bodies sufficient strength to facilitate handling.

Ball milling

The comminution of the whiskers was achieved using a ball mill consisting of a polyurethane cylinder (to avoid contamination) and sintered silicon nitride grinding media (again to avoid unnecessary contamination of the mixture). A technique for estimating the aspect ratio of the whiskers was identified by Milewski [1]. This technique enables the milling time for the reduction of the whisker length to a given aspect ratio to be estimated. The optimum condition was found to be milling 100 g batches for 4 hours.

Homogeneous blending of the reinforcing whisker phase with the matrix material using a blast mill

The homogeneous blending of the whiskers and powders could not be achieved by turbular mixing and ball milling. A patented NEL technique [2] which had been successful used for the mixing of alumina fibres in an aluminium matrix was modified for the blending of the silicon carbide whiskers in the silicon nitride matrix. The basic equipment, Figure 1, comprises a vibratory unit which feeds the material to a container via a fine sieve. The sieve unit rotates generating a centrifugal force which pushes the material through the sieve, Figure 2. The shearing action of this movement simultaneously assists the breaking down of hard agglomerates within the starting powders, and mixes the whiskers and powder homogeneously, Figure 3.

Production of free flowing powders for easy mould or die filling

The main difficulty in processing ceramic powders arises from the small particle size. The small particles have a high free surface energy. This results in a high thermodynamic driving force to reduce this surface energy by bonding together. While this bonding is advantageous in the sintering process it is undesirable but not in the preconsolidation stage as the particles tend to "stick" together, and to the walls of the dies during pressing.

The obvious solution is to increase the size of the powder particles. This has a detrimental effect on the sintered density of the product, unless more sintering additives are added, but this would then have an adverse effect on the high temperature properties of the material. The solution appeared to be to generate artificial soft agglomerates. Agglomeration of monolithic ceramics are usually achieved via a spray drying process. This technique is costly and produces relatively small and hard agglomerates. An easier, more flexible and

Figure 1. Photograph of the centrifugal milling equipment.

Figure 2. Photograph of the inside of the high speed mixing equipment showing the screen and rotor device which rotates at high speed and which the powders and whiskers move through.

Figure 3. Scanning electron micrograph of the homogeneous distribution of silicon nitride powder and silicon carbide whiskers, after high speed mixing.

cheaper technique was thus identified. This involved the compaction of quantities of the mixture in an isostatic press (35 MPa) followed by sieving of the resulting mass into various size fractions — 150/212, 212/300, 300/425 µm.

Consolidation of the agglomerated powders and whiskers

The free flowing agglomerated mixture was uniaxially pressed using a 25·4 mm diameter die set. A range of pressures were applied (15, 30, 45, 52, 60 and 75 MPa) and assessed in conjunction with the different agglomerate sizes to optimise green density values for this specimen size (approximately 3 mm thick). This particular size and shape of specimen was chosen to enable comparative strength analysis of the composite materials to be undertaken without having to produce expensive four point flexural specimens. Specimens were subsequently isostatically pressed (207 MPa), to improve green density levels and to produce a material with isotropic properties (both in terms of removal of density variations and preferred whisker orientation which results from the uniaxial pressing stage).

Sintering

Specimens were sintering in a Jones Brothers overpressure (10 Bar) sintering furnace. The furnace has graphite elements with thermocouple control to 900°C and twin beam radiation pyrometer control from 900 to 2300°C. The sintering variables explored were:

— The powder bed.
— The sintering atmosphere.
— The heating rate at each stage.
— The overpressure at each stage.
— The time at a given temperature.
— The rate and stage at which depressurisation takes place.

The organic binder was burnt off at 500°C for 1 hour, various different combinations of heating rate, and application of overpressure and sintering times were studied. The most consistent set of results came from the following schedule:

(1) Ramp rate = 10°C/min to 500°C, 1 hour dwell.
(2) Ramp rate = 30°C/min to 1650°C (6 bar pressure at 1400°C).
(3) Ramp rate = 10°C/min to 1700°C, 2 hour dwell, 10 bar pressure).

Although this sintering schedule did not yield the best possible results in terms of achievable strength levels, it provided consistent reproducible results (all specimens sintered with no signs of decomposition) which were used to determine the effectiveness of the processing variables which was the purpose of this study.

Analysis of specimens

The disc specimens were tested using a biaxial test jig. A minimal amount of machining was required to produce the specimens (the composite discs were diamond ground to ensure that the faces were flat and parallel. This technique was found to give comparable results with that of four point flexural testing (providing volume of material under test is considered) at a fraction of the cost:

$$\sigma_r = \sigma_t = 3W/\pi t^3((1 + v)\ln(R_s/R_l) + (1 + v)(R_s^2 - R_l^2)/2R_d^2 \quad (1)$$

where:

σ_r = stress in a radial direction
σ_t = stress in a tangential direction
W = load in Newtons
t = thickness (less than 2·5 mm)
v = Poisson's ratio
R_l = radius of load ring circle (10 mm)
R_s = radius of load support circle (19·24 mm)
R_d = radius of disc being tested

SEM analysis was carried out on the resulting fracture faces to determine the mode of failure and to study the resulting microstructure of the composites.

3. RESULTS AND DISCUSSION

3.1 Factors influencing the preparation of silicon nitride/silicon carbide composites

In general the starting materials have a great influence on the final sintered composite. The particle size of the starting powder, for example, affects the green specimen preparation and the densification process. The fineness of the powder affects the sinterability, the high free surface energies of the small particles act as a driving mechanism by bonding together to reduce this free surface energy. Silicon nitride has a surface layer of silica which facilitates the liquid phase sintering process. The smaller particles have a higher specific surface area hence a larger volume of liquid can be produced during sintering. This makes for a faster rearrangement stage during the sintering process. However, the purity of the starting powder is as important as the size of the particles. The presence of impurities could result in poor high temperature properties, since both Fe and Al impurities form silicide phases which have very low melting points. Carbon may react with any oxygen present in the specimens; the volatile CO gas produced results in porosity which effectively reduces density values. Pores can also act as stress concentrators having an adverse effect on the strength properties.

Effects of powder/whisker packing

One of the most important factors in composite fabrication is the achievement of a good green density. The green density of the specimens will be poor unless a good packing of particles is achieved. If powder particles are perfect spheres and a perfect ABAB-type packing arrangement can be achieved, this results in 30% voidage in the body. If particles the diameter of the voids are introduced, the voidage will be reduced to 26%. It is, therefore, advantageous to have a particle size distribution in the powders. The morphology of the starting powders is therefore of prime importance. In addition to the reconstructive polymorphic transformation of alpha to beta silicon nitride as one of the driving forces for densification, alpha silicon nitride is favoured as a starting powder as the grains are equiaxed which is beneficial for good particle packing. The final beta grains have an acicular morphology which in itself is a toughening mechanism.

Packing problems are further complicated by the addition of acicular shaped whisker reinforcements. Many of the commercially available whiskers have very high aspect ratios (approximately 200). Not only does this cause problems in terms of maximising green densities, but also in properties. The long whiskers tend to bundle together producing voids in the sintered body and anisotropic properties.

Milewski [1] found that fibres with aspect ratios of 20-40 yielded the best result and that a noticeable reduction in properties resulted in materials where aspect ratios of fibres exceeded 75. A technique was developed which enabled aspect ratios of fibres to be estimated. This technique was used to assess the milling time required for the reduction of the silicon carbide whiskers to an acceptable aspect ratio. The method involved the measurement of a bulk volume of whiskers after various milling times, translating this to a relative bulk volume and then using graphic relationship developed by Milewski [1], Figure 4, to estimate the aspect ratio. It was found that milling the whiskers for more than a 4 hour period yielded a very minimal additional reduction in

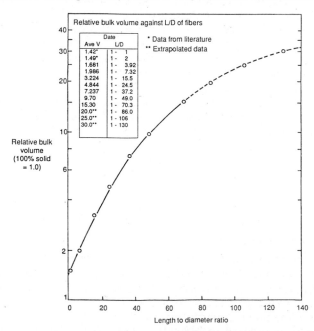

Figure 4. Packing curve for fibres with varied L/D ratios, Milewski, (1988).

whisker length. It was found that the addition of the silicon nitride powders prior to milling had little effect on the milling time required (100 g batch milled for 4 hours). The addition of the liquid binder at the milling stage helped consolidate the fine ceramic dust, which was both a processing and a safety advantage. Turbular blending, although not producing a homogeneous mix, was advantageous in the successful milling of the whiskers.

Effects of blast milling

The major constraint in the successful production of randomly orientated short fibre CMCs is that of ensuring a random homogeneous distribution of fibres in the matrix. Conventional techniques have involved a wet mixing stage, this leads to extra processing stages (drying and breaking down of the resulting cakes) which not only increased the processing costs, but increases the possibility of introducing defects into the material.

The high speed mixing or blast mill technique, Figures 1 and 2, has been shown to provide a successful route in the dry mixing of ceramic powders and whiskers, Figure 3. The centrifugal acton which forces the material through the fine mesh in shearing action helps to break down hard agglomerates in the powder which can act as inclusions in the final sintered component. Work carried out by Kendall [3] examined the effects of hard agglomerates in ceramic bodies using three different methods of compacting the titania powder. The three methods chosen were pressing, slip casting and plastic forming. The mean defect size c and the spread of defect sizes were calculated from Weibull plots. The worst results were obtained from the mildest forming method — slip casting, which gave the largest defects and the lowest reliability. Pressing gave smaller defects because it applies higher compaction pressures. Plastic mixing which generates large shear forces gave a further improvement in reliability. All three methods resulted in relatively large defects compared to the size of the grains implying that some of the hard agglomerates are very difficult to break down. It was discovered that these agglomerates could be broken down further by applying high shear forces to the powder in a polymer solution and extruding the mix through a small orifice. The hard agglomerates from the starting powders within the synthesised agglomerates were subjected to a similar shearing process during the high speed milling process to assist in the break down (in the range of the sieve size).

The defects can result from the smaller agglomerates within the larger agglomerates as well as inclusions or trapped gas bubbles. These small agglomerates which survive the compaction process were the strength limiting flaws in the compacts. Thus, it follows that the strength of the powder compacts is influenced by the harshness of the compaction process. Compaction methods which apply high stresses will break down the small agglomerates and so should produce stronger assemblies.

Effect of soft agglomerates on processing

The fine nature of the silicon nitride powders and silicon carbide whiskers makes them difficult to handle; due to high free surface energies they tend to

Figure 5. Scanning electron micrograph of 300/212 μm agglomerates of silicon nitride containing no silicon carbide whiskers showing crushed agglomerates.

"stick" together or to the die walls during uniaxial pressing. Additionally, safety considerations have to be made when handling silicon carbide whiskers due to the aspect ratio of submicron diameter whiskers. Work by Kendall, described above, indicates that hard agglomerates in the powder can be detrimental to the properties of the sintered body. It was discovered that by producing soft agglomerates, whose fracture strength was less than that of the uniaxial pressing load, Figure 5, easily handleable material could be produced, which promoted improved efficiency.

The pressure at which the agglomerates were pressed was chosen specifically to ensure that they would be broken down when subsequently uniaxially pressed. Figure 5 is an SEM micrograph of the soft agglomerates that have broken during analysis. These soft agglomerates were produced by isostatically pressing at 35 MPa; if this pressure was increased to 70 MPa the agglomerates would require more energy to break down during uniaxial pressing. It was also noted that there was a maximum uniaxial pressure for the production of the specimens, so that the crushing point of the agglomerates could be exceeded to maximise the green density. Exceeding this pressure was a waste of energy as no further densification could be achieved, and lateral cracking of the specimen would result. The critical level was approximately 60 MPa for this size of specimen.

The morphology of the agglomerates varied with size, Figure 6, which shows agglomerates of sizes: 212/150 μm, 300/212 μm and 425/300 μm produced by Cold Isostatic Pressing (CIP) at 35 MPa. Figure 7 shows the variation in the density levels with varying agglomerate size and with the addition of silicon carbide whiskers. The green density increases with decreasing agglomerate size. It was noted that the addition of silicon carbide whiskers to the matrix resulted in the reduction of the green density. This is due to the bridging effect of the whiskers which adversely affects packing of the particles. Further compaction by CIP increased green density values, more

(a) 212/150 μm   (b) 300/212 μm   (c) 425/300 μm

Figure 6. Different sized soft agglomerates:

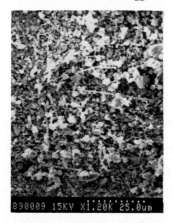

Figure 7. Scanning electron micrograph of 425/300 μm agglomerates of silicon nitride containing 10 wt% silicon carbide whiskers showing the surface of an agglomerate and the powder and whisker distribution and packing within the agglomerate.

noticeably for smaller agglomerates. It was also noted that the homogeneous distribution of the whiskers was unaffected by the agglomeration stage, as shown in Figure 8 which is an SEM micrograph of agglomerates of size 425/300 μm containing silicon carbide whiskers.

3.2 Consolidation

The effects on density of the different size fractions of agglomerates produced were assessed in terms of uniaxial pressing loads and the results are given in Figure 9. The optimum value for the size of specimen was 60 MPa. The density is obviously dependent on the compaction ratio — the finer the powder the higher the ratio required to achieve maximum density.

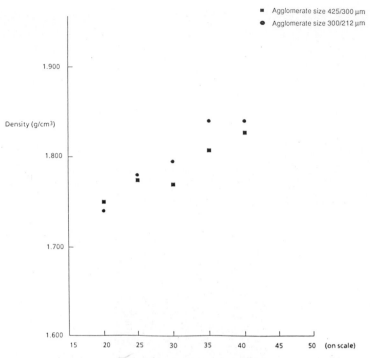

Figure 8. Graph showing the variation in density with uniaxial pressure.

It was found that pressures greater than 60 MPa resulted in lateral cracking of the specimen. After uniaxial pressing the specimens were isostatically pressed to both improve green density levels and to remove the preferred orientation of the whiskers resulting from the uniaxial pressing stage. The uniaxial pressing also results in a non-uniform density distribution throughout the specimen, again leading to property anisotropy within the sintered component as well as distortion from uneven shrinkage.

The material was uniaxially pressed and then isostatically pressed, rather than directly isopressed to enable discs to be produced for strength analysis using a biaxial testing jig (which would not involve a great deal of costly diamond machining as would the preparation of four point flexural specimens). The specimens were pressed at 207 MPa, higher levels showed no significant improvement, particularly when considering the additional time and energy requirements for increased pressures. The decompression rate was also found to be critical — slow rates of compression were required in order to prevent cracking of the specimens (as the mould exerts tensile forces on the specimen).

For near net shape production, material could be uniaxially pressed to provide some green strength, then subsequently green machined and isostatically pressed. Alternatively a mould of the desired component shape

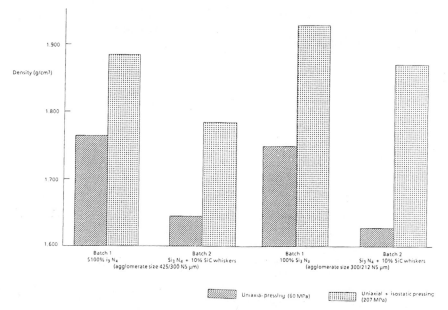

Figure 9. Graph showing the variation in densities with material, consolidation technique and agglomerate size.

could be produced for direct isostatic pressing. Design of moulds is critical to ensure correct allowances are made for shrinkage during densification, and this will obviously vary with volume of reinforcing phase.

3.3 Sintering

The successful densification of compacted green powder bodies involves the removal of pores between the starting particles accompanied by the shrinkage of the body with particles growing together to form a strong bond with adjacent particles. Sintering only occurs if a mechanism for material transport and a source of energy to activate and maintain material transport are present. The primary mechanisms for material transport are diffusion and viscous flow. Diffusion of nitrogen in these highly covalent materials is very slow making solid state sintering virtually impossible. Therefore, in the sintering of silicon nitride a liquid phase sintering technique is used. The composites under study are intended for low stress, load bearing applications, where high temperature strength is not of primary importance. Thus, magnesia was chosen as the sintering additive. Magnesia reacts with the surface silica to form a liquid oxynitride glassy phase (the formation temperature depending on the relative proportions of silica and magnesia in the system). The liquid phase sintering of silicon nitride occurs in three stages:

— Particle rearrangement
— Solution-diffusion-precipitation
— Coalescence

Magnesia forms a low viscosity liquid phase, making densification relatively fast and alpha to beta silicon nitride transformation easier.

Properties of silicon nitride and silicon carbide make them ideal candidates for composite materials. Unlike the β-sialon/silicon carbide system, Bower [4], it has been found that no interaction occurs at the interface between the reinforcing phase and the matrix during sintering.

When producing a multiphase ceramic material it is necessary to determine the compatibility of the materials, and the conditions (temperature, pressure and composition of the gas phase) at which the densification of the composite can take place without decomposition of the matrix or the reinforcing phase by the reactions:

$$Si_3N_4 \rightarrow 3Si + 2N_2 \tag{2}$$

$$3SiC + 2N_2 \rightarrow Si_3N_4 + 3C \tag{3}$$

Work undertaken at the Max-Planck Institute [5], demonstrated that there is a critical region in which both the matrix and the reinforcing phase will stay stable during overpressure sintering. Sintering furnace overpressure effectively increase the possible sintering temperature (using Le Chateliers principle the reaction can be moved to inhibit dissociation of Si_3N_4, Equation (2)). This provides a mechanism for faster sintering and hence implies improved efficiency in terms of speed and energy.

The silicon carbide may react with the nitrogen under overpressure sintering conditions. The critical partial pressure of nitrogen required to follow reaction (3) can be calculated:

$$\Delta G(reaction) = -RTlnk$$

$$\Delta G(reaction) = \Delta G(Si_3N_4) - 3\Delta G(SiC)$$

$$k = a(Si_3N_4) \cdot a(C)/a(SiC) \cdot p^2(N_2)$$

Using available data [6] and assuming the activity of Si_3N_4 and carbon is unity, the nitrogen pressure at a given temperature is shown in Figure 10. This figure shows that increasing the sintering temperature requires an increase in nitrogen pressure to follow reaction (3), increasingly rapidly as the temperature exceeds 1550°C. Below 1550°C it is difficult for reaction (3) to occur as the free energy of the reaction (3) (ΔG_{r3}) is almost a negative value (considering errors in the thermodynamic data), as shown in Figure 11. Figure 10 shows that nitrogen overpressures of above 9 atm. at a sintering temperatures of 1700°C could result in the start of reaction (3). A 10 bar overpressure was used at a sintering temperature of 1700°C, this could be a reason for the reduction in sintered density values when the percentage of silicon carbide whiskers in the matrix is increased.

3.4 Powder beds

Various powder beds were assessed for sintering the composites:

(a) 100% boron nitride
(b) 50% boron nitride/50% silicon nitride
(c) 60% boron nitride/30% silicon nitride/10% magnesia

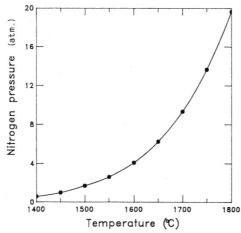

Figure 10. Graph showing the relationship between nitrogen pressure and temperature.

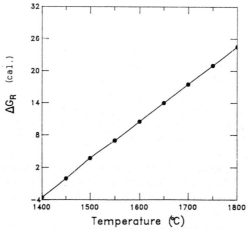

Figure 11. Graph showing the relationship between ΔG(reaction) and temperature.

The powder beds are used to prevent the decomposition of the silicon nitride specimens. Boron nitride is used as the decomposition temperature is considerably higher than that of silicon nitride and provides a layer between the crucible and the specimen. Embedding the specimens in a powder bed suppresses the decomposition and vaporisation by establishing the equilibrium partial pressures over the constituents. The use of the same additive concentration in the powder bed as in the specimen helps to prevent concentration gradients, across the specimen and the powder bed. This type of bed has been found to be unsuccessful as the powder bed starts to sinter. The best results were achieved by placing the specimens in bed type (b). This is due to the provision of both nitrogen and silicon vapour around the specimen which assists in preventing the volatilisation from the specimen.

3.5 X-ray analysis

X-ray analysis indicated that complete alpha to beta transformation had occurred during the sintering process and that no secondary crystalline phases were present, only the glassy oxy-nitride phase at the grain boundaries. No reaction between the SiC and Si_3N_4 appears to have occurred (no interfacial reacted zones were visible on the SEM micrographs). It is possible that a reaction between the SiO_2 and the SiC whiskers could have taken place at low heating rates, below the expected eutectic temperature, resulting in the formation of CO gas. This would then affect the eutectic temperature as the proportion of SiO_2 in relation to the sintering additive would be reduced. Therefore, density levels could possibly be adversely affected if very low heating rates were used.

A reaction between the matrix and the silicon carbide may have occurred, Equation (3), which would have resulted in the formation of silicon nitride and carbon deposits both of which would not be detected as additional phases during X-ray analysis.

3.6 Results of strength and Weibull analysis

The nature of the study involved the recording of changing process variables. The best method of monitoring the results was to assess these changes in terms of strength and Weibull modulus parameters (a statistical measure of the reliability of the material) and SEM analysis of fracture faces.

The Weibull modulus is a measure of the probability of the specimen to fail by a flaw or defect in a given volume of a material. This analysis, in basic terms, gives a value which is inversely proportional to the standard deviation of the test results, *i.e.* the larger the scatter the lower the Weibull modulus and the higher the probability of the specimen to fail. The result is more accurate with a large number of specimens, though it should be noted that any unusually low values should be investigated before the data is included in the series of results. The calculations can account for differences in volumes of material tested in different types of testing procedures.

Conventional techniques involve flexural testing of specimens, which are produced by diamond machining to close tolerances and are therefore expensive. For the purposes of this work a comparative measure was required, using a technique which was relatively cheap (as a large number of tests were required).

The results of the strength analysis and the Weibull analysis are listed below:

(1) 0% addition of silicon carbide whiskers, sintered at 1700°C for 2 hours:

Mean strength (212/150 µm)	= 527·9 MPa
Weibull modulus	= 4·2
Mean strength (300/212 µm)	= 442·8 MPa
Weibull modulus	= 2·8

(2) 10% addition of silicon carbide whiskers, sintered at 1700°C for 2 hours:

Mean strength (300/212 μm)	= 556·2 MPa
Weibull modulus	= 7·2
Mean strength (425/300 μm)	= 469·9 MPa
Weibull modulus	= 4·5

(3) 20% addition of silicon carbide whiskers, sintered at 1700°C for 2 hours:

Mean strength (300/212 μm)	= 199·9 MPa
Weibull modulus	= 6·4

The results demonstrate that the smaller agglomerate size results in a higher comparative strength value, and a higher Weibull modulus, Weibull modulus values increase with increasing whisker additions.

The sintered density values also have an effect on the strength values. It was noted that some of the specimens with relatively low bulk densities had higher strength values than those with higher bulk densities. SEM analysis showed that this was an effect of the size of the porosity (the pores act as defects in the material) it appears that the porosity size has a more dominant effect on the strength of the material.

4. CONCLUSIONS

A practical method for the production of randomly orientated short fibre near net shape CMC components has been identified. A blending process has been developed to homogeneously mix SiC whiskers with Si_3N_4 powder, the resulting soft agglomerates providing a mechanism for easy handling of fine powders (with consequent benefits in safety, processing time and mechanical properties). Smaller soft agglomerate sizes yielded the best results, in terms of strength and Weibull Modulus. Bulk densities correlated directly with strength values, and addition of increasing whisker loadings yielded higher Weibull Modulus values. N_2 overpressure sintering reduced the potential for decomposition of Si_3N_4 at higher sintering temperatures, but reactions with the reinforcing phase need to be considered.

ACKNOWLEDGMENTS

Discussions with, and help from Dr. R. J. Dower of NEL are gratefully acknowledged.

REFERENCES

1. MILEWSKI, J., *Adv. Ceram. Materials*, **1**, 36, (1986).
2. National Engineering Laboratory publication No.: NEL/446/316, "Method of Mixing Short Fibres with Metallic Powder," UK appln. No.: 8500685, (1988).
3. KENDALL, K., *Powder Met.*, **31**, 28, (1988).
4. BOWER, R. M., Ph.D. Thesis, University of Strathclyde, 1990.
5. Powder Metallurgy Laboratory, Max-Planck Institute, Stuttgart, **20**, 26, (1988).
6. TURKDOGAN, T. E., *Physical Chemistry of High Temperature Technology*, Academic Press, New York, 20, (1980).

Sol-Gel Fabrication of CMCs

R. S. RUSSELL-FLOYD, R. G. COOKE, B. HARRIS, J. LAURIE,
R. W. JONES, T. H. WANG and F. W. HAMMETT
School of Materials Science, University of Bath, Bath, BA2 7AY

1. INTRODUCTION

The wider commercial application of CMCs to general engineering applications is currently limited by their inherently high cost and difficulty in fabrication to net size and shape. In an attempt to address these limitations, the School of Materials Science at the University of Bath working in collaboration with Ceramic Developments (Midlands) Ltd. have developed a sol-gel processing route for the fabrication of complex, low-cost, ceramic matrix composite components. Based on freeze gelation of aqueous colloidal sols and adaptations of fibre reinforced plastics technology for component manufacture, this processing route is outlined below followed by a consideration of some of the most important processing parameters which affect CMC properties.

2. THE PROCESSING ROUTE

To an aqueous colloidal silica sol, containing typically 15–50% by weight silica of particle size 5–125 nm is added a ceramic filler powder typically 0·5–20 μm in particle size. This filled sol is then combined with the appropriate reinforcing fibres and a component formed by either filament winding or hand lay-up for continuous fibres or by casting or injection moulding for short discontinuous fibre reinforcement. The filled sol is then cooled by rapid freezing, typically using liquid nitrogen as the refrigerant. Nucleation and growth of ice crystals causes concentration of the colloidal particles into a densely packed gel network. The ice crystals, typically 1–10 μm in diameter, are the main source of porosity in the gel, and consequently their removal during thawing and drying results in relatively low capillary stresses and low shrinkages (less than 1% by volume with suitable fillers). Subsequent sintering yields a rigid CMC component but with volume fractions of porosity ranging from 0·3–0·8 depending on colloidal silica, filler and fibre content. Although certain applications such as filtration devices or air bearings might make use of such porosity, mechanical properties are enhanced through porosity reduction by repeated liquid phase infiltration and resintering.

3. CRITICAL PARAMETERS IN SOL-GEL PROCESSING

Although there are major differences in the products of sol-gel derived CMCs produced by different fabrication routes (filament winding, casting and hand lay-up) and with different reinforcing fibres, these differences are largely the result of differences in the ability of the fibres to transfer their properties in the composite. Two broad classes of CMC can therefore be defined. Those containing relatively low volume fractions of short fibres in which the matrix properties and in particular porosity dominates composite properties. And those containing higher volume fractions of continuous fibres in which fibre degradation controls composite performance. In both cases, critical matrix parameters for sol-gel processing are those associated with (a) the colloidal silica sol, (b) the filler, (c) the influence of the specific process — winding or casting, (d) the gelation rate, (e) the sintering regime, and (f) the infiltration/densification process.

The main selection criteria for the reinforcing fibres has, for the purposes of economic manufacture of CMCs, been cost. Such fibres as Saffil and carbon therefore become very attractive having an order of magnitude advantage over the more commonly used silicon carbide based fibres.

For a silica sol to be capable of freeze gelation, it appears that it must contain unaggregated colloidal silica particles in an aqueous medium. The viscosity and solids content of the sol are of particular importance. A low viscosity sol is advantageous for infiltration between fibres, but a high solids content decreases initial matrix porosity and therefore increases green matrix strength. The solids content is governed both by the weight fraction of silica in the colloidal silica sol and by the type and proportion of filler powders added. Not only does an increased solids content increase the sol viscosity, but it also decreases the fibre volume fraction achievable, both disadvantages.

Process variables associated with the manner of freeze gelation, for example, cooling rate also play an important role. The growth of the ice crystals which ultimately generate the inherent porosity within the sol-gel matrices shows similarities to the growth of grains in the casting of metals. As shown by Figure 1, a layer of fine equiaxed porosity at the surface gives way to longer elongated pores caused by columnar crystal growth in the direction of heat flow during cooling. In general, however, the higher the cooling rate the smaller and more equiaxed the porosity.

The combination of sintering temperature and the type of filler powder used are key variables in determining the properties of continuous fibre CMCs. Generally speaking, fillers can be grouped into two categories: *reactive* fillers which are typically glasses or glass ceramics which act as sintering aids to the *non-reactive* fillers which are the more refractory ceramics such as silica and alumina. It has become evident that reactive fillers are usually inappropriate for use with the oxide ceramic fibres because they cause too strong a fibre/matrix interface bond and degrade the fibres particularly at higher sintering temperatures. This leads to weak, lower toughness composites. Composites containing the non-oxide ceramic fibre, carbon, can gain from the improved matrix strength and higher interlaminar shear strength facilitated by the presence of a reactive filler. This point is demonstrated by the flexural stress/strain curves included as Figure 2. The two sets of data are from similar CMCs except that to one an addition of a lithium alumino silicate glass

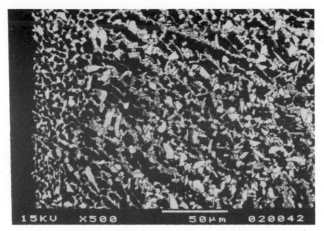

Figure 1. A typical microstructure of a freeze gelled silica sol filled with a 1–10 μm mullite filler. The sol has frozen from left to right.

Carbon fibre reinforced CMCs tested in three point flexure

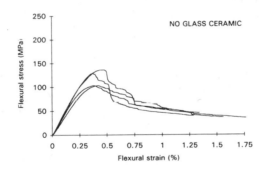

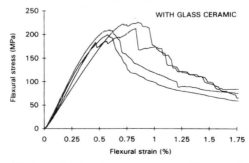

Figure 2. Typical sets of flexural stress/strain curves for unidirectional carbon fibre reinforced freeze-gelled CMCs manufactured by hand lay-up, demonstrating the effect of using a reactive filler (glass ceramic).

ceramic has been added, causing an obvious improvement in both strength and, from the areas under the curves, toughness.

The densification of the CMCs is conducted by repeated submersion in a colloidal silica sol of the finest primary particle size available (5–9 nm), followed by re-sintering. This can reduce the volume fraction of residual porosity to around 0·2.

4. CONCLUSIONS

Optimisation of the various process variables associated with the freeze-gelation route to CMCs is a complex problem because of the large number of variables and their propensity to interact. The two most important for long fibre reinforced are the filler and sintering regime, whilst for short fibre reinforcement, those variables which reduce porosity are most important.

Ceramic Materials Based on Alumina-Chromia-Mullite Compositions

H. G. EMBLEM, T. J. DAVIES, A. HARABI and V. TSANTZALOU

Manchester Materials Science Centre, University of Manchester/UMIST, Grosvenor Street, Manchester, M1 7HS, UK

ABSTRACT

Developing on our previous work on the solid-state chemistry of alumina-chrome refractories, the system Al_2O_3-Cr_2O_3-SiO_2 was studied. The results showed that in ethyl-silicate-bonded alumina-chrome refractories the silica formed from the ethyl silicate reacts with the aluminium (III) oxide to form mullite, the aluminium (III) oxide and the chromium (III) oxide form a solid solution which accounts for the improved resistance to slag attack and damage by thermal shock. This preliminary result led to a study of the sub-system mullite-Al_2O_3-Cr_2O_3. A composition range for a fine grain suitable for forming engineering ceramics or the matrix of a refractory shape was established. This grain is formed by comminuting a solid solution of chromium (III) oxide in aluminium (III) oxide with fine mullite grain. Engineering ceramics or refractory shapes may be made from the grain by conventional forming methods. The refractory shapes retain their strength at high temperatures. Compositions up to 10 wt% chromium (III) oxide are preferred (with about 2.5 wt% chromium (III) oxide being optimum), with up to 10 wt% fused mullite. This gives good densification and good strength, which is retained at high temperature. With 10 wt% fused mullite, grain size and intergranular porosity are small.

1. INTRODUCTION

Incorporating chromium (III) oxide in alumina refractory bodies improves resistance to slag attack and thermal shock damage. The chromium (III) oxide may be added as a chrome ore. For example, slag attack may be reduced by adding [1] fine chrome to coarse alumina. In mullite-chrome systems an iron-chromite ore can be the source [2] of chrome. Incorporating fine chromium (III) oxide in ethyl-silicate-bonded alumina or mullite refractories [3] improves resistance to damage by thermal shock and slag attack; in this case chromium (III) oxide is preferred to a chrome ore. Examples of refractory shapes made from ethyl-silicate-bonded alumina-chrome include nozzles and well-blocks for sliding-gate systems, bubble plugs and burner liners.

The present paper summarises and develops on our earlier work [4–9] on alumina-chrome refractories. Preparative procedures, grain growth and densification were studied in sintered compacts prepared from aluminium (III) oxide/chromium (III) oxide mixtures. XRD studies confirmed solid solution formation showing also that the corundum lattice dimensions contracted with 7 wt% chromium (III) oxide, other compositions giving lattice expansion. The good densification and uniform grain growth in sintered bodies of mixes containing 7 or 14 wt% chromium (III) oxide are related to the high resistance to thermal shock damage and slag attack shown by ethyl-silicate-bonded alumina-chrome refractories. Developing on these observations, some diffusion couple studies of Al_2O_3-Cr_2O_3-SiO_2 compositions, as models for ethyl-silicate-bonded alumina-chrome refractories, are reported. Because mullite was always formed in the diffusion zone, the sub-system mullite-Al_2O_3-Cr_2O_3 was studied, resulting in the establishment of a composition range for a fine grain suitable for forming engineering ceramics or the matrix of a refractory shape.

2. EXPERIMENTAL

The materials and experimental procedures have been described in detail previously [4–9]. Table 1 summarises the properties of the materials and Table 2((a) and (b)) summarises milling and blending procedures, and the preparation routes for various mixes. Compacts were prepared by compacting the powder mix at 310 MPa in a tungsten carbide lined steel die, then sintered at varying temperatures for varying times and cooled in the furnace. Densities were determined before and after sintering by weighing and measuring. Grain size and grain structure were determined by optical or scanning electron microscopy [5, 6, 8, 9]. The effect of milling and blending was determined by X-ray diffractometry and the corundum lattice constants calculated [6]. The strength of some compacts was determined [10] by diametral compression.

Diffusion couples were prepared from AMS9 aluminium (III) oxide, M100 chromium (III) oxide and BDH precipitated silica. They were prepared by compacting (~ 1 MPa) and pre-sintering (1450°C); data on tablets of powders blended are given in Table 2; two tablets placed face-to-face formed a single diffusion couple (Figure 1(a)). The couple was then heated to the required temperature in the range 1600–1800°C, at a heating rate of 20°C min^{-1} and held at temperature for 15, 25, 45 or 60 minutes. Procedures for microscopic examination (optical and/or electron) of the couple after heating, also X-ray diffraction procedures were as previously described [6, 7, 9].

Table 1. Properties of starting materials

Material	Source	Particle size	Impurities
Aluminium (III) oxide	MA95 (BA Chemicals Ltd.)	Average 4 μm angular	Na_2O — 0·55% min. SiO_2 — 0·07% max.
Aluminium (III) oxide	A17 reactive (Alcoa)	Average 2 μm round	
Aluminium (III) oxide	AMS9 (Sumitomo Industries)	Average 0·55 μm	
Chromium (III) oxide	M100 (British Chrome & Chemicals	Average 0·4 μm	Soluble impurities 1% max
Silica	Precipitated silica (BDH Chemicals Ltd.)		Al_2O_3 — 0·75% $K_2O + Na_2O$ ~ 0·5%
Mullite	'Fused' (Keith Ceramic Materials Ltd.)	Average 1·4 μm	Glass 3·2% Na_2O ~ 0·25%

Table 2(a). Milling and blending procedures

Method	Description
High energy milling (hem)	Using a chrome steel pot containing hardened steel balls, for MA95 and A17 reactive. Vibrate at high frequency, gives highly deformed particles. For AMS9, use zirconia balls.
Ball milling (bm)	Using chrome steel balls in a mill of hardened rubber. Wear and abrasion gives a round particle shape, used only with MA95 and A17 reactive.
Tungsten carbide milling (tcm)	Using tungsten carbide mortar and pestle, mechanically driven. Used only with MA95 and A17 reactive. The action is a mixture of shear fracture, abrasive wear and particle rolling.
Milling procedure	For MA95 and A17 reactive. Mill for 2 h (hem, bm or tcm) to break down particle aggregates.
Blending procedure	For MA95 and A17 reactive, after milling add M100 andblend mixture for 4 h (hem, bm or tcm). For AM59 no milling, blend mixture with M100 hem for 6 minutes up to 120 h. Powders for diffusion couple studies prepared by hem. Mullite-Al_2O_3/Cr_2O_3 solid solution grain mixes prepared by hem.

Table 2(b). Preparation routes for alumina-chromia-mullite mixes

Aluminium (III) oxide + 2·5 wt% chromium (III) oxide powders

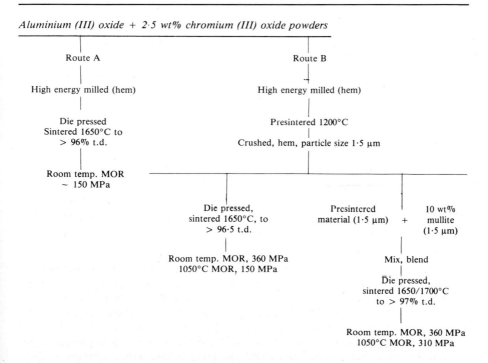

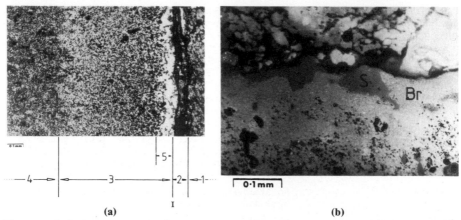

(a)   (b)
                          I

Figure 1. Diffusion couples, (a) General view of typical diffusion couple 1 — silica rich zone;
2 — cracks; 3 — diffusion zone; 4 — alumina rich zone; I — interface; 5 — Bright
phase (Br in (b)).
(b) Diffusion couple after heating, showing bright etching phase (Br) consisting of small
crystals of mullite and corundum, S is a silica rich phase.

Solid solutions of AMS9 aluminium (III) oxide and M100 chromium (III) oxide for use in the study of the sub-system mullite-Al_2O_3-Cr_2O_3 were prepared as previously described [5, 9]. The fused mullite grain has been described previously [11]. Milling and blending procedures are given in Table 2. Compacts were prepared as described above, sintered at varying temperatures for varying times and cooled in the furnace. The properties of the compacts were determined as described above. The strength determinations at ambient temperature were determined by diametral compression [10]. Strength determinations at temperature were determined by three-point blending [6].

3. RESULTS AND DISCUSSION

Aluminium (III) oxide and chromium (III) have a similar rhombohedral structure. The system aluminum (III) oxide-chromium (III) oxide is a simple binary system, forming solid solutions of chromium (III) oxide and aluminium (III) oxide. Our XRD results [4, 6, 7] confirm solid solution formation. Small amounts of chromium (III) oxide give a red colour, larger amounts giving a green colour. The colour difference could relate to the energies of the ligand field splitting [12].

3.1 Powder preparation

Milling is essential to reduce the large primary particle size of the MA95 material. With both MA95 and A17 reactive materials, milling and subsequent powder blending each give [6, 7] lattice distortion and strain. The observations are summarised in Table 3; it was noted that the chromium (III) oxide content improved powder blending and homogenisation. This is further improved by using AMS9 aluminium (III) oxide, which has a particle size comparable with that of M100 chromium (III) oxide. High energy milling

Table 3. Densification, grain growth and microstructure observed in sintered aluminium (III) oxide-chromium (III) oxide compacts. Compaction at 310 MPa

Aluminium (III) oxide material	Chromium (III) oxide/wt%	Observations
MA95 or A17 reactive	3	Fine-grained porous microstructure at 0·6–1 h sintering time. Densificaton and formation of large grains by selective grain growth at 8 h sintering time.
MA95 or A17 reactive	7 or 14	Grain growth is more uniform and equiaxial. MA95 gives low density, consistent with loss of chromium by volatilisation.
AMS9	2·5	Preferred quantity used for diffusion couple studies and for sub-system mullite-Al_2O_3-Cr_2O_3. This quantity minimises possibility of loss of chromium by volatilisation.
AMS9	2·5–5 or more	Decreasing densification of compacts sintered at 1600°C for 1 h or less. Increasing sintering temperature and time increases density.
AMS9	10 or 15	Grain structure changes from polygonal (in AMS9) to tabular.

(hem) is the preferred milling and blending procedure. With AMS9 material and M100 chromium (III) oxide, corundum lattice distortion and strain are small.

Powders to be used in preparing diffusion couples were also hem-blended for various times before forming into tablets. Powder mixes used in the study of the sub-system mullite-Al_2O_3-Cr_2O_3 were prepared by hem blending of fused mullite grain and the appropriate solid solution of aluminium (III) oxide and chromium (III) oxide.

3.2 Densification, microstructure and grain growth

For MA95 and A17 reactive materials, 14 wt% chromium (III) oxide gives compacts having the highest 'green' density. Increasing the chromium (III) oxide content improves powder blending and increases the 'green' density of the compacts, because the chromium (III) oxide particles fill the interstices between the MA95 or A17 reactive alumina particles. Increasing the compaction pressure increases the 'green' density as would be expected. In general a small aluminium (III) oxide particle size (as in A17 reactive (hem) and AMS9) gives densification at low sintering temperature, also rapid grain growth. The powders were compacted at 310 MPa, with no binder, permanent or fugitive, being added as it was considered that these might alter the microstructure. No attempt was made to optimise the particle size distribution before compaction. The results [6, 7] are summarised in Table 3. The low

Figure 2. Microstructure of compact from powder mix 7 wt% Cr_2O_3; 93 wt% A17 reactive Al_2O_3 (2 h tcm), blended 4 h hem. Sintered at 1600°C for 8 h. Density 3895 kg m^{-3}, grain size 9 μm.

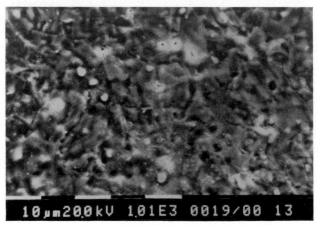

Figure 3. Microstructure of compact from powder mix 14 wt% Cr_2O_3; 86 wt% A17 reactive Al_2O_3 (2 h tcm), blended 4 h hem. Sintered at 1600°C for 8 h. Density 3860 kg m^{-3}, grain size 8 μm.

density observed [6] in some sintered compacts prepared from MA95 material could be due to loss of chromium. This is consistent with the larger particle size of MA95 material and suggests that volatilisation is more rapid than solid solution formation. When heated above 1000°C in air or oxygen, chromium (III) oxide volatilises, due to the reaction

$$Cr_2O_3 \text{ (s)} + 3/2 \text{ } O_2 \text{ (g)} \rightarrow 2CrO_3 \text{ (g)} \tag{1}$$

Figures 2 and 3 show the microstructure of compacts prepared from A17 reactive aluminium (III) oxide. It is considered that these microstructures explain the resistance to thermal shock damage and slag attack found in ethyl-silicate-bonded alumina-chrome refractories. Micrographs of other compacts are given in references [6] and [8].

The particle sizes of AMS9 and M100 materials are similar. Compacts sintered at 1600°C or 1650°C have a smaller grain size compared to AMS9 material, although at 1700°C there is some grain growth. For compacts containing 15 wt% chromium (III) oxide, the grain size is smaller at all temperatures and times of sintering. The change in grain size can be described [7, 9] by a surface diffusion controlled and pore drag mechanism for almost all compositions. However, compositions in the range 5–7 wt% chromium (III) oxide show anomalous behaviour.

3.3 Diffusion couple studies

AMS9 aluminium (III) oxide was chosen as the material to be used in diffusion couple studies because it is similar in particle size to M100 chromium (III) oxide. The chromium (III) oxide content of each tablet was limited to and fixed at 2·5 wt% to avoid the possibility of loss of chromium by volatilisation [5, 9].

A typical diffusion zone is shown in Figure 1(b). The bright phase was resolved at high magnification into a fine structure of two crystalline phases, suggesting that it is a devitrified glass. Relating the chemical composition of the bright (glassy) phase to the Al_2O_3-Cr_2O_3-SiO_2 phase diagram [13] identified the crystals as mullite and corundum. This was confirmed by XRD studies.

XRD study and microscopic examination of all the diffusion couples sintered at various temperatures for varying times showed that:–

(a) mullite was always found in the diffusion zone, the chromium (III) oxide content being roughly constant at 2·8 wt%;

(b) segregation of chromium (III) oxide occurred in silica-rich regions, aluminium (III) oxide-chromium (III) oxide solid solutions being formed in the other part of the couple;

(c) chromium (III) oxide solubility in the silica phase was low.

Figure 4 shows a typical distribution in a diffusion couple. The results suggest that in ethyl-silicate-bonded alumina-chrome refractories the silica formed from the ethyl silicate reacts with the alumina 'fines' to form mullite, which could contain about 2·8 wt% chromium (III) oxide. The major part of the chromium (III) oxide forms an aluminium (III) oxide — chromium (III) oxide solid solution. This will account for the observed improved resistance to thermal shock damage and to slag attack.

3.4 The sub-system mullite-Al_2O_3-Cr_2O_3

Based on the results of the diffusion couple studies, the sub-system mullite-Al_2O_3-Cr_2O_3 was examined. Powders were prepared by comminuting alumina-chromia solid solutions with fine fused mullite grain. Fused mullite, rather than sintered mullite was chosen because of the high proportion of crystalline mullite (97·5%) and the relatively low glass content (3%). Up to 25 wt% fused

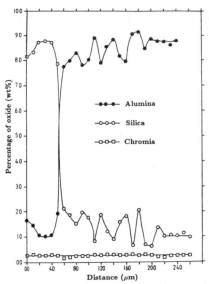

Figure 4. Concentration profiles of alumina, silica and chromia across the diffusion zone of a couple sintered at 1600°C for 60 minutes.

mullite was used, with an alumina-chromia solid solution containing 2·5 wt% chromium (III) oxide.

The effect of sintering conditions on grain growth and densification as the weight percentage of fused mullite is increased has been described previously [5, 9]. Up to 10 wt% fused mullite results in tabular grain formation with low intergranular porosity, further addition of fused mullite (up to 25 wt%) gives a duplex microstructure. Densification is improved by the addition of fused mullite.

Tensile strength determinations [5, 9] at ambient temperature showed that the addition of 2·5 wt% chromium (III) oxide to AMS9 aluminium (III) oxide doubled the strength of compacts sintered at 1600°C for 15 minutes. Further addition of chromium (III) oxide, up to 5 wt%, decreased the strength gradually and large amounts gave very low strengths. When the temperature and time of sintering were increased, the strength decreased. The results suggest that the strength is controlled mainly by the density when sintered, the fraction of open porosity and the grain size after sintering. Compacts prepared from a powder mix obtained when a solid solution of AMS9 aluminium (III) oxide and M100 chromium (III) oxide (2·5 wt%) was comminuted with 2·5 wt% fused mullite showed [5, 9], when sintered at 1600 or 1650°C for times up to 60 minutes, an increase in tensile strength at ambient temperature, in comparison with compacts prepared from a powder mix comprising AMS9 aluminium (III) oxide and M100 chromium (III) oxide (2·5 wt%). Comminuting with greater amounts of fused mullite decreased the strength, but the decrease was less when the time of sintering was increased to 4 h. However, on comminuting with 10 wt% fused mullite, there was a significant

Table 4. Flexural strength at temperature of test pieces prepared from solid solution of AMS9 aluminium (III) oxide and chromium (III) oxide (2·5 wt%), also from grain mix obtained by comminuting the solid solution of AMS9 aluminium (III) oxide and chromium (III) oxide (2·5 wt%) with fused mullite grain (10 wt%)

	Flexural strength (MPa) of test piece	
Temperature of test (°C)	Prepared from solid solution	Prepared from grain mix
850	360	360
1050	150	310
1500	140	160
Ambient	350	360

increase in strength when the time of sintering was increased to 12 h.

Table 4 gives the flexural strength at temperature of test bars comprising the solid solution of AMS9 aluminium (III) oxide and M100 chromium (III) oxide (2·5 wt%) and the flexural strength at temperature of test bars prepared from the grain mix obtained by comminuting the solid solution of AMS9 aluminium (III) oxide and M100 chromium (III) oxide (2·5 wt%) with 10 wt% fused mullite. Each series of test bars was sintered at 1650°C for 4 h, then tested at temperature (three-point-bending). The results show that the bars made from the grain mix obtained by comminuting the solid solution of AMS9 aluminium (III) oxide and M100 chromium (III) oxide with fused mullite (10 wt%) have better retention of strength at temperature. This grain mix may be shaped by uniaxial or isostatic pressing. It could also be used as a matrix material in the production of refractory shapes.

4. SUMMARY AND CONCLUSIONS

(a) The preferred method of making aluminium (III) oxide-chromium (III) oxide mixes to be used in producing refractory shapes is high energy milling, first to break down clusters of primary particles, then to blend in chromium (III) oxide.

(b) A study of diffusion couples comprising Al_2O_3, Cr_2O_3 and SiO_2, Cr_2O_3 suggests that the formation of mullite and an Al_2O_3-Cr_2O_3 solid solution explains the improved resistance to thermal shock and slag attack observed in ethyl-silicate-bonded alumina-chrome refractories.

(c) Comminuting a solid solution of aluminium (III) oxide and chromium (III) oxide with fused mullite gives a grain mix from which refractory shapes having good retention of strength at high temperature can be prepared.

REFERENCES

1. Kyushu Taika-Renga Kabushiki Kaisha, British Patent 1, 533, 890.
2. C. Taylor's Sons Company, British Patent 1, 421, 418.
3. SHAW, R. D. & SHAW, C., British Patent 1, 313 498.
4. DAVIES, T. J., EMBLEM, H. G., NWOBODO, C. S., OGWU, A. A. & TSANTZALOU, V., *Polyhedron,* **8,** 1765, (1989).

5. HARABI, A. & DAVIES, T. J., in Euro-Ceramics, **2**, Eds. G. de With, R. A. Terpsera and R. Metselaar, publ. Elsevier Applied Science, London, **2.576**, (1989).

6. DAVIES, T. J., EMBLEM, H. G., NWOBODO, C. S., OGWU, A. A. & TSANTZALOU, V., *J. Mater. Sci.,* **26**, 1061, (1991).

7. EMBLEM, H. G., DAVIES, T. J., HARABI, A. & TSANTZALOU, V., in *UNITECR '91 Congress Preprints,* 291, (1991).

8. TSANTZALOU, V., M.Sc. Thesis, UMIST, (1984).

9. HARABI, A., Ph.D. Thesis, UMIST, (1990).

10. OVRI, J. E. O. & DAVIES, T. J., *Mat. Sci. & Engineering,* **96**, 109, (1987).

11. DAVIES, T. J., EMBLEM, H. G., JONES, K., MOHD. ABD. RAHMAN, M. A., SARGEANT, G. K. & WAKEFIELD, R., *Br. Ceram. Trans. J.,* **89**, 44, (1990).

12. HARTFORD, W. H., in *Kirk-Othmer Encyclopedia of Chemical Technology, 3rd Edn.,* **6**, publ. John Wiley & Sons, Inc., New York, **87**, (1979).

13. ROEDER, P. L., GLASSER, F. P. & OSBORN, E. F., *J. Amer. Ceram. Soc.,* **51**, 585, (1968).

High-Temperature Advanced Ceramic Coatings for Carbon-Carbon Fibre Composites

*DR. ALEXANDER A. POPOV and †DR. MICHAIL M. GASIK
*UkrNIISpetsstahl, 330600 Zaporozhye, Ukraine
†Institute of Materials Technology, Helsinki University of Technology,
SF-02150 ESPOO, Finland
(Visiting Research Scientist, Dnepropetrovsk Metallurgical Institute, Ukraine)

ABSTRACT
Advanced oxide ceramic coatings have been developed for use on carbon-carbon fibre composites to protect them from oxidation during long-term application at working temperatures of up to 1350–1400°C, and during a range of thermal cycling. These coatings have a high strength adhesion to the carbon fibre substrate and have a matched thermal expansion coefficient, low porosity and high thermal stability as well as good thermal shock resistance. During the present sequence of testing no reaction between carbon composite and coating has been observed in the studied temperature range and so the coatings look promising for high-performance applications.

1. INTRODUCTION

Carbon-carbon fibre reinforced composites (CCFRC) are widely used in different areas of engineering due to their unique combination of properties, in particular high mechanical strength together with low density. The major advances which have been made during the last 10 to 15 years are related to high temperature applications in aircraft and space industry. However, the potential applications of CCFRC are limited by the high affinity of carbon fibres and matrix to oxygen. This reaction leads to a maximum service temperature of 400–450°C to avoid oxidation runaway of the material. It has been reported elsewhere that the most common technology for CCFRC production is as shown in Figure 1. As a rule, the density of the woven material before impregnation is 40–45% of theoretical. Two kinds of CCFRC textures used industrially are "Novoltex" and "Skinex" [1]. Novoltex has medium mechanical properties and is used mainly for thick articles (missile nozzles, heat-protective shields, and so on). Skinex has a higher level of properties and has been designed especially for thin rigid shells, resistant to thermal and mechanical loading.

To protect CCFRC's in oxidising environments at high temperatures different coatings have been developed. For example, SEP has developed the CVD [1] method for silicon carbide layer formation. This material was reported to have a good toughness, shear resistance, high adhesion, and low thermal expansion coefficient. But at the same time it has been noted to contain a lot of microcracks, generated during material preparation due to the different linear expansion coefficients of carbon fibre and SiC. That is why this kind of coating has found application for short-term service only (engine nozzles, outlet tubes in solid fuel missile carriers; single-use thermal protective plates for HERMES space shuttle; etc. [1]).

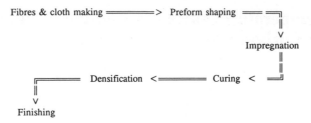

Figure 1. Conventional technology of CCFRC production using resin impregnation.

Another example of a higher performance coating has been given by Chou *et al.* [2] who have studied ZrC/SiC and $ZrSi_2$ based coatings on 3D woven graphite/graphite (C-C) made by infiltration and reaction bonding techniques. Their key point was in a special post-treatment carried out for enhancement of the coating by sealing the microcracks of the coating. However, despite its successful application in air at 1500°C for a long time without weight loss, it is not suitable during thermal cycling as microcracking results from the incompatible thermal expansion of matrix and coating. In addition, these coatings are considered to be unable to protect CCFRC in oxygen-rich environments.

The aim of the present paper is to report upon the development of a multilayer [3] coating which seems to obviate most of the difficulties associated with the coating of CCFRC.

2. EXPERIMENTAL

2.1 Materials

Typical properties of commercial composites [1] and one used in the present study (CCPM-1) are shown in Table 1. The properties of all C-C materials are in accordance with ones published by Sociéte Europeéne de Propulsion (SEP), and it is believed that they have been measured for particular direction of the fibres or weaves. Furthermore, they have relatively high densities, as a result

Table 1. Properties of different CCFRC and graphite

Material or its trade name	Density, g/cm^3	Ultimate tensile strength, MPa		Linear expansion coefficient,* $\cdot 10^{-6} K^{-1}$
		20°C	2000°C	
2D C-C Laminate	1·60	200	—	2·8
C-C Novoltex	1·75	45–80	—	5–7
C-C 4D	1·90	20–80	—	5–8
C-C Skinex	1·75	200	—	—
CCPM-1	1·45	390	480	3·5
Graphite	1·9–2·1	40	—	—

*Depending upon the direction of the fibres

of several stages of pyrodensification. In contrast the material CCPM-1 has been densified in a single stage operation.

Today, there is a good understanding of the correct choice of fibres and cloth direction, impregnation, curing, and densification of composites. As a result the final properties of all known commercial examples of CCFRC are not very different from each other. More important questions appear when one tries to use these composites for the production of specific parts.

2.2 Coatings

The coatings used in the present work are based on the concept of a silicon containing coating which is sealed beneath a silica glass layer at high temperatures. Such a coating configuration leads to a self-heating effect in the case of microcracking, much as in the case of molybdenum disilicide heating elements.

From a consideration of thermochemical stability such oxide phases are the only reliable ones for extended usage in oxidising conditions. However, if an oxide layer is deposited directly on to the carbon fibre surface, then reduction of the oxide may occur as the temperature is increased. Even in situations where little reduction occurs there may still be problems with mismatch of thermal expansion of the components.

Consequently, it was decided to develop a laminate coating structure which contains different sub-layers of controlled anisotropy. No sharp interfaces exist between the layers and the separation of oxide layer from direct contact with the carbon fibre avoids chemical reduction reactions. The data presented in subsequent sections are for a proprietary zirconia-alumina based mix applied on to a silicon sub-layer on CCPM-1 composite.

3. RESULTS AND DISCUSSION

Several oxides and their mixes have been studied, and zirconia-alumina based mixes are believed to be the most promising. The following properties (Table 2) have been obtained for ZrO_2-Al_2O_3 based coatings on CCPM-1 composite. It is evident that the combination of properties listed is rather good for high performance applications. One of the most important characteristics, oxidation resistance during high temperature thermal cycling, has been studied by conventional and high-temperature continuous thermogravimetry. No significant weight loss or gain was observed. The tensile strength of the coated composite after oxidation under conditions of thermal cycling remains at the initial value and implies that this material could be considered as a substitute for C/SiC composites (Table 3).

The low-temperature cyclic oxidation testing of this material was also carried out. This testing was performed under conditions where ordinary mechanisms of coating regeneration, i.e. reformation of protective silica layers, do not work. The samples had 200 μm average thickness of ZA-coating and were tested in flowing air for 20 thermal cycles of 20°C to 600°C in 1 hour followed by a 1 hour cool to 20°C. The final weight loss of the samples did not exceed 10% relative to initial weight. Typical SEM microstructures of the

Table 2. Properties of the zirconia-alumina (ZA) based coating on the CCPM-1 composite

Properties	Values
Coating material density, g/cm³	2·3
Closed porosity, %	2
Coating thickness, mm	0·2
Adhesion strength "coating/carbon," MPa	4·0
Linear thermal expansion coefficient, 10^{-6} K^{-1}	3·3
Number of thermal cycles tested in air (20 → 1350 → 20°C/2 min)	50
Thermal shock resistance after 50 cycles	No changes

Table 3. Comparison between the properties of ZA-coated CCPM-1 composite and C/SiC composite, made by SEP

Properties	C/SiC composite [1]	ZA coated CCPM-1
U.T.S. at 20°C, MPa	200–350	370–400
Oxidation temperature, °C	1550	1350
Holding cycle time, min	20	20
Total holding time, h	27	200
U.T.S. reducing ratio, %	20	0

specimens after such thermal cycling are shown in Figure 2 for different magnifications. It can be seen that there is no oxidation under the coating, and the coating itself has not been separated during fracture trials. Such encouraging observations require more studies to be made to determine the probable mechanism of oxidation resistance under these conditions and to study the extended service life of such coatings under a variety of oxidising conditions.

4. CONCLUSIONS

The results reported briefly above show a very promising possibility of obtaining high performance oxide coatings for CCFRC which will be suitable for heavy duty conditions in aircraft and space applications.

In this context it is considered useful to determine the possibilities of using these kinds of coatings for the following purposes:

1. Nozzles, outlet pipes and pipelines of solid fuel jet propulsion units, where high-temperature thermal shock in oxygen environments occur.
2. Thermal protection shields of a space shuttle in places of high loading, where there may exist the possibility of re-use on many occasions.
3. Using the coatings for high-temperature joining of CCFRC, in particular for large articles with a complex shape.

Figure 2. Microstructures of fracture surfaces of coated samples after thermal cycling in air flow.

These types of applications are the subject of present work which will also include studies of adhesion strength during the application of mechanical loads, cyclic fatigue levels, and microstructure changes at the interface and of the surface of fibres during oxidation and thermal cycling. Once comprehensive evaluation data have been obtained they will be published elsewhere.

REFERENCES

1. LACOMBE, A. & BONNET, C., *Ceramic matrix composites. Key materials for future space plane technologies,* AIAA Pap., No. 5208, 1, (1990).
2. CHOU, S. T., CHOU, H. Y., WU, H. D., *et al., Surface protection of 3D C/C composite at elevated temperature, poster presentations at Mat. Tech. '90, Symp. Mater & Proc.,* June 10–18, 1990, Espoo, Finland.

Coating of Single Crystal Alumina Fibres with Zirconia Sols

P. M. BROWNIE, C. B. PONTON and P. M. MARQUIS

Interdisciplinary Research Centre in Materials for High Performance Applications and School of Materials and Metallurgy, The University of Birmingham, Edgbaston, Birmingham, B15 2TT, UK

ABSTRACT

Oxide fibre-reinforced oxide matrix composite ceramics appear to be the materials for high temperature applications in oxidising environments. Ceramics are inherently brittle and in order to optimize the toughening mechanisms, which arise due to fibre reinforcement, the interface between the fibre and the matrix has to be relatively weak. It is often essential to coat the fibre to control the properties of the interface. In this work single crystal alumina fibres have been coated by electrostatic attraction during immersion in a zirconia sol. An electrophoresis apparatus has been adapted to enable the electrostatic interaction between sol particles and a fibre to be quantified.

1. INTRODUCTION

Metal alloys are limited to operating temperatures of around 1000°C. Therefore, high temperature (> 1000°C) applications, such as the interior of combustion chambers, require the use of ceramics. Although borides, carbides and nitrides have the highest possible operating temperatures, oxides have the advantage of not decomposing under oxidising conditions such as those experienced in engines. Thus, unlike non-oxide materials oxide ceramics require no protective coating to prevent decomposition and hence, are the obvious choice for high temperature applications in oxidising environments.

Ceramic monoliths are inherently brittle and to prevent catastrophic failure a reinforcing phase, which both strengthens and toughens the ceramic, has to be introduced. Extensive work has been carried out over the last 30 years into the reinforcement of ceramics and it has been shown that fibre reinforcement offers a maximization in toughness and strength [1]. The stronger the reinforcing fibres are, the stronger the composite will be. Thus, the maximization of fibre strength has progressed to the use of single crystal fibres as opposed to polycrystalline fibres. Single crystal fibres are inherently stronger than the equivalent polycrystalline fibres because there is no possibility of grain growth or grain boundary slippage at elevated temperatures [2]. The maximum strength of a fibre reinforced composite is achieved when the interfacial bonding between the fibre and the matrix is strong enough to allow effective load transfer between the two. Unfortunately the toughening mechanisms, such as crack deflection and fibre debonding, bridging and pull-out, depend on the establishment of a weak fibre/matrix interface [3]. Although this weak interface may occur naturally in a system, it is usually necessary to coat the reinforcing fibres to establish the desired characteristics of the interface.

The process used to coat the fibres is required to be as simple, cheap and effective as possible, allowing for the all round continuous coating of fibres. These criteria eliminate several well established coating methods such as physical vapour deposition and chemical vapour deposition. The most obvious choice is a sol coating method whereby the coating is deposited on the fibres by electrostatic attraction. A full understanding of the deposition process is essential to allow modelling of prototype systems and to enable the prediction of the viability and effectiveness of sol coating in other systems. Thus, the aims of this work are to establish a coating system which can be used in the production of oxide fibre-oxide matrix reinforced composites and to develop a novel technique, based on electrophoresis, to directly quantify the interaction between a sol particle and a fibre.

2. EXPERIMENTAL RESULTS AND DISCUSSION

2.1 Sol Immersion Coating of Fibres

Single crystal α alumina fibres obtained from Saphikon Inc. were used as substrates for fibre coating. These fibres are continuous, approximately 130 μm in diameter, which have been produced by continuous edge defined film fed growth. The zirconia sols used in this study were produced by hydrothermally processing a zirconium acetate precursor to produce monoclinic zirconia particles 0·8 μm in diameter on average.

A zirconia sol was used to set up a range of sols of differing pH. The pH of the sol was varied by the addition of 0·1 M $NaOH_{(aq)}$. Lengths of cleaned and mounted Saphikon single crystal fibres were then placed in test-tubes containing the sols. The orientation of the fibres was not predetermined but was usually found to be horizontal. The fibres were left for 24 hours before rinsing with distilled water and drying in a laminar air flow cabinet. Energy dispersive X-ray analysis (EDX) of the coatings was undertaken on a JEOL 5200 scanning electron microscope (SEM). The normalized results obtained are shown in Figure 1. Although the EDX results are not fully quantitative, the general trend of the results shows that from no detectable zirconia coating at pH 3, the extent of the coating increases up to a maximum around pH 6 and then falls away towards the higher pH values.

A series of micrographs taken on a Hitachi 4000-S Field Emission Gun SEM support the EDX analyses by showing that at low pH very few zirconia particles adhere to the surface but as the pH is increased the mobility of the zirconia sol falls and the apparent particle density increases to a maximum around the isoelectric point, with fewer but larger particles adhering after flocculation has occurred (see Figures 2-5).

Even with the most dense coating there still appear to be gaps between the particles. This is probably an optical phenomenon, since it is impossible to tell, especially at extremely high magnifications, whether the dark background is the fibre surface or a lower layer of particles. This is supported by the fact that when the fibres were snapped and viewed end on, the coating was definitely observed to be continuous in the case of the fibre coated at pH 5·7 (see

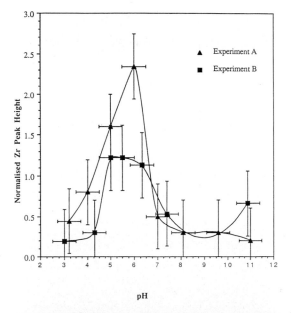

Figure 1. Normalized Zr EDX peak from fibre as a function of pH of immersion sol.

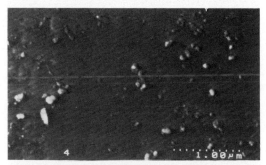

Figure 2. Saphikon fibre immersed in zirconia sol at pH 3·3.

Figure 3. Saphikon fibre immersed in zirconia sol at pH 4·6.

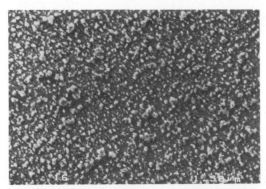

Figure 4. Saphikon fibre immersed in zirconia sol at pH 6·1.

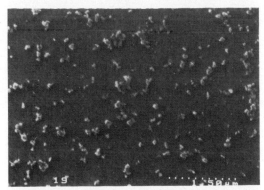

Figure 5. Saphikon fibre immersed in zirconia sol at pH 7·4.

Figure 6. Saphikon fibre immersed in zirconia sol at pH 5·7. Fracture surface of fibre viewed end on.

Figure 6) and may also be continuous at pH 4·7, 5·2, and 6·1. Examination of the fibre coated at pH 10 showed different coating density distributions in different regions of the fibre. This is thought to be because the fibre was placed horizontally in a flocced solution. Thus, even though the sample was agitated, the uppermost surface of the fibre would undergo more collisions with the flocced sol particles than the lowermost surface, due to the action of gravity making the particles fall through the solvent.

Comparison of the mobilities of a zirconia sol and an alumina sol, which it was hoped would be representative of the surface charge behaviour of the alumina fibre, could lead to a prediction of the interactions between the fibre and the sol particles (see Figure 7). Although the mobility values of the alumina sol give an indication of the polarity and relative magnitudes of the effective surface charge of the fibre as the pH is changed, it will not give an indication of the relative magnitudes between the fibre and sol particles. Thus, at low pH the potential difference is minimized and hence the attraction between particle and fibre is minimized, reducing the number of collisions.

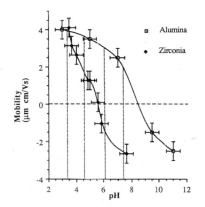

Figure 7. Mobility of sol particles as a function of pH. The dotted vertical lines indicate the pH of Figures 2–5.

The electrostatic potential barriers, around both the fibre and the sol particles are maximized at the extremes of pH so that the probability of a collision resulting in a particle adhering to the fibre surface is minimized. Hence, very few particles are observed on the surface at either very low or very high pH. As the pH increases in the range pH 4–6 the potential difference will increase. This will have the effect of moving particles towards the fibre because they will tend to move down a potential gradient. This mechanism is only applicable in a concentrated sol where the repulsion from surrounding particles is greater than the electrostatic repulsion provided by the fibre. The potential barrier surrounding the fibre is reduced, hence there are more collisions with a greater probability of adhesion, which results in progressively thicker coatings. In the region of the isoelectric point pH of the zirconia sol (pH ~ 5·7 ± 0·5) the electrostatic component of the potential barrier will be effectively removed but for coating, the pH has to be accurately measured and maintained. The fact that at the isoelectric point (iep) a very small change in ionic concentration brings about a very large change in pH makes coating at the iep a theoretical rather than practical consideration. At a pH just greater than the isoelectric point ($\sim$ 7) flocculation will be slow and electrostatic attraction will occur between the fibre and the now larger particles. Thus, the collision probability will be greatly increased until all the surface charge of the fibre has been screened off from the bulk of the sol by the oppositely charged deposition material.

2.2 Electrophoresis and its Use in Investigating Fibre-Particle Interactions

The Rank Brothers Mark II Particle Micro-electrophoresis apparatus is designed to measure the velocity of charged particles under a known electric field and from these measurements enables the calculation of the mobility and zeta potential of the particles. A standard Mark II electrophoresis apparatus was adapted by redesigning the sample cell to include a fibre mount such that a fibre could be suspended in the centre of the cell. The modified cell with fibre is shown schematically in Figure 8. The aim of this adaptation was to enable fibre-particle interactions to be studied directly with the possibility of measuring the charge density of the fibre.

The apparatus essentially consists of a light source, focusing and magnifying optics and the cell system with temperature and voltage control (see Figure 9). Light is focussed into a parallel beam which is shone through a variable slit and a filter, which blocks the red end of the spectrum to reduce problems with radiant heating. The light then passes through a low power dark field condenser, the purpose of which is to spread the light into an annular configuration. This "ring" of light is focussed in the observation plane of the cell so that none of the incident light is directly observed but only that light which is forward scattered from the particles. The particles in the plane of focus are thus observed as pin-points of light (see Figure 10).

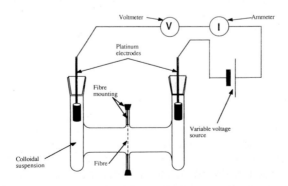

Figure 8. Adapted electrophoresis cell.

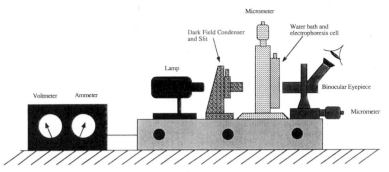

Figure 9. Schematic representation of electrophoresis apparatus.

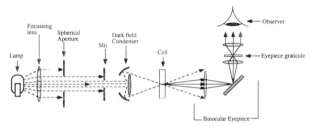

Figure 10. Optical path of light scattered in electrophoresis cell.

With a fibre placed in the cell, the motion of several particles was monitored moving both towards and away from the fibre by changing the polarity of the electrodes. Thus, a velocity plot can be determined for the case when the field due to the fibre and the field due to the electrodes are in opposition and when they are additive. The results are shown in Figure 11. It can be clearly seen from the magnitude of the comparable velocities and the acceleration of the particles that the field due to the particle has a noticeable effect up to > 200 μm from the fibre surface. If there were no electrostatic interaction between the particle and the fibre then the velocity profile for approach and retreat from the fibre would be symmetrical. This clearly is not the case.

When an external field is applied to the particles in addition to the field applied by the fibre several forces have to be taken into account other than simply the electrostatic attraction. Firstly, there is the viscous force which is proportional to the particle velocity and acts to oppose any movement through the fluid. Secondly, there are the forces involved in the fluid dynamics of flow past a cylinder. The fluid flow past the fibre in this configuration can be shown to be laminar from calculations of the Reynolds number of the cell channel and the fibre.

The laminar flow of the fluid past the fibre can be represented by a series of flow lines, between which the rate of flow is constant (see Figure 12). This means that as the fluid flows past the fibre it will be deviated causing a change in pressure, the fluid closest to the fibre will flow the fastest and the flow lines

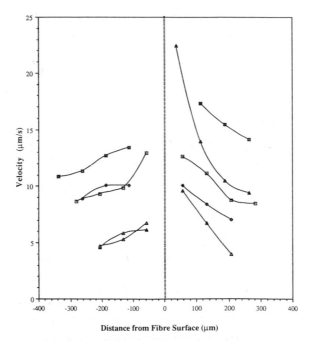

Distance from Fibre Surface (μm)

Figure 11. Velocity of charged particles versus distance from fibre. Positive distances are representative of the cases where the field from the fibre and the external field act in the same direction and negative distances are where the fields are in opposition.

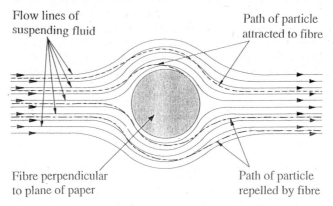

Figure 12. Particle motion past a fibre. Repulsive and attractive particles are shown crossing the flow lines as they move past the fibre.

will be closer together at this point. If a particle is attracted or repelled by the fibre whilst approaching or flowing past the fibre then the particle will cross the flow lines and change both its inherent position in the cell and its velocity away from the fibre. The velocity changes because of the parabolic flow profile across the cell which is due to electro-osmotic effects [4]. Thus, if the particle is moved towards and away from a charged fibre the velocity profile will change with each change in direction. If the fibre had no electrostatic charge then the velocity profile of the particle would remain the same regardless of the number of times the particle flowed past the fibre. (This disregards the effect of Brownian motion moving the particle laterally).

The change in particle velocity simply due to fibre particle electrostatic interaction and the change in particle velocity due to crossing flow lines is impossible to distinguish without analysing the pattern of fluid flow past the fibre in considerable detail. A first approximation can be made to the force exerted on the particle by assuming no viscous resistance and assuming that the particles do not cross flow lines, which could increase or decrease the particle velocity. Thus, from purely electrostatic considerations, the field experienced by a particle due to the fibre can be estimated from the following equation.

From Coulombs Law the field around an infinitely long fibre is defined as

$$E_f = \frac{r\sigma}{\varepsilon_0 x}$$

for $x > r$

where E_f is the field strength, ε_0 is the permittivity of free space, r is the fibre radius, σ is the surface charge per unit area and x is the distance from the centre of the fibre. Thus, the field is inversely proportional to the distance from the surface of the fibre. This approximation is reasonably good because the length of the fibre is very much greater than the distance of the particle from the fibre surface.

A sol of known mobility u was used and hence from the equation

$$E = \frac{u}{v}$$

where E is the electric field strength experienced by the particle and v is the experimental velocity, the resultant field at different distances from the fibre surface can be calculated. Since the velocity of each sample particle was measured approaching and retreating from the fibre the effect of the external field and fluid flow can be eliminated by halving the sum of the velocities at the corresponding distances. The results are plotted in Figure 13. By fitting a straight line to the data it can be seen that all the sets of data are close to a perfect linear fit and closely agree with the linear inverse relationship of field with distance from the fibre. In fact, the particle for which results were taken closest to the fibre surface gave a gradient of 1·000. Given the preliminary nature of the results, the accuracy and linearity of the results is highly encouraging.

3. CONCLUSIONS

The aim of this work was to lay down coatings in as simple and economical a way as possible, so that in future by altering the coating material or the deposition parameters the fibre/matrix interface can be optimized to enhance toughening without reducing the strength of the composite to an unacceptable

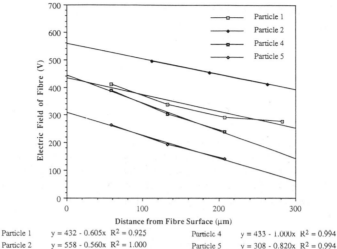

Particle 1 $y = 432 - 0.605x$ $R^2 = 0.925$ Particle 4 $y = 433 - 1.000x$ $R^2 = 0.994$
Particle 2 $y = 558 - 0.560x$ $R^2 = 1.000$ Particle 5 $y = 308 - 0.820x$ $R^2 = 0.994$

Figure 13. Fibre electric field versus distance from fibre surface.

degree. The method of coating chosen for initial investigation was an immersion electrostatic coating because of the simple method of processing, the low costs involved and the fact that it is one of the few coating methods which can be scaled up to accommodate all round coating of continuous fibres or the coating of woven fibre preforms. The coating procedure was carried out by immersing the fibres in a sol and allowing the sol particles to coat the fibres by electrostatic attraction. By varying the pH of the sol it has been shown that the depth of the coating has been maximized close to the isoelectric point pH of the sol. Hence, coating by immersion has proved successful in producing thin coatings. How these can be improved and the optimum thickness for inclusion in a fibre matrix composite has yet to be investigated.

Since the fibres are single crystals, it cannot be assumed that the surface will behave in the same way as polycrystalline particles and hence a method of predicting and directly measuring the interaction between the two was sought. A viability study into the use of an adapted electrophoresis cell to monitor and quantify the attraction between sol particles and a fibre has proved successful. It was shown that there is a measurable electrostatic interaction between sol particles and a fibre even without taking into account the fluid flow patterns and viscous retardation effects. This is very encouraging because this adapted microelectrophoresis technique is the only method being developed to measure the charge density and hence zeta potential of individual fibres. Thus, this approach will be pursued and refined through further modelling and calibration.

REFERENCES

1. SAMBELL, R. A. J., *et al., J. Mat. Sci.,* **7,** 676, (1972).
2. HAGGERTY, J. S., *Nasa CR-120948 AD1 73235,* (1972).
3. RUHLE, M. & EVANS, A. G., *Prog. Mat. Sci.,* **33,** 85, (1989).
4. HUNTER, R. J., *Zeta Potential in Colloid Science,* Academic Press, 150, (1981).

Co-precipitated Zirconia-Hafnia Powders

H. P. LI, J. WANG* and R. STEVENS
Division of Ceramics, School of Materials, The University of Leeds, Leeds, LS2 9JT
*IRC in Materials for High Performance Applications,
The University of Birmingham, Birmingham, B15 2TT

ABSTRACT

High performance transformation toughening is obtainable in ceramic matrices containing zirconia-hafnia solid solution inclusions. Zirconia-hafnia powders of fine particle size and narrow particle size distribution are the important starting materials for these high temperature transformation-toughened ceramics. On the one hand, zirconia-hafnia powders of nanosizes can be prepared via the co-precipitation route using mixed zirconium and hafnium nitrate salts as the starting materials. On the other hand, the characteristics of the resultant oxide powder are affected strongly by the processing parameters. Zirconium-hafnium hydroxide is formed when zirconium and hafnium oxynitrates are co-precipitated in an ammonia solution at pH 10 to pH 11. The characteristics of the calcined zirconia-hafnia powders, such as crystallite size, particle size, particle size distribution, and the degree of powder agglomeration, are largely dependent on the way in which the co-precipitated hydroxides are dried. A comparison is made of three different drying routes, including organic solvent dehydration, microwave drying, and conventional IR heating lamp drying, on the basis of powder characteristics investigated using XRD, BET surface area, DTA, TGA, sedigraph, SEM, and TEM. The powders processed using these three different drying routes were also assessed in terms of their sinterability at 1300°C.

1. INTRODUCTION

Transformation toughening has recently been well established in zirconia-dispersed ceramics, following the discovery of stress-induced transformation toughening in the mid 1970's [1]. However, zirconia toughened ceramics do not exhibit the desirable mechanical properties at elevated temperatures, simply because of the low monoclinic to tetragonal transformation temperature occurring in zirconia [2, 3]. Although hafnia has been widely suggested as a potential substitute for zirconia for high temperature applications, the high temperature toughening has not been well demonstrated in hafnia dispersed ceramic matrices [4, 5]. It is difficult to retain a metastable tetragonal hafnia phase in a ceramic matrix, as the critical particle size for the spontaneous tetragonal to monoclinic transformation in hafnia is in the range of 5 to 10 nm [6, 7]. Due to their strikingly similar crystal structures, hafnia and zirconia form continuous solid solution over the entire range of composition [8]. The physical characteristics of zirconia-hafnia solid solutions, such as the monoclinic to tetragonal transformation temperature, and the critical grain size for the spontaneous tetragonal to monoclinic transformation, are dependent on composition. Specifically, all these parameters obey approximately the rule of mixtures. Naturally, zirconia-hafnia solid solution offer a technologically important toughening mechanism for engineering applications in the intermediate temperature range. However, zirconia-hafnia solid solutions should have a finer inclusion size than monolithic zirconia, in order to retain the metastable tetragonal phase in the ceramic matrix.

Co-precipitation is the most commonly used method for producing doped zirconia and hafnia powders [9, 10]. Hydroxides are formed by adding zirconium or hafnium salts such as the nitrate or chloride in an ammonia solution, followed by drying and calcination to dehydrate the hydrous oxides. This involves the crystallization of amorphous zirconia or hafnia and subsequent crystal growth, together with possible particle agglomeration when the co-precipitated gel is dried and calcined. An appropriate drying procedure is of paramount importance, in order to avoid undesirable crystal growth and powder agglomeration. Zirconia or hafnia powders with a coarse crystallite size and a high degree of powder agglomeration exhibit a low sinterability and therefore a high sintering temperature [11]. Furthermore, it may be thermodynamically impossible to retain the metastable tetragonal phase in a sintered ceramic matrix, when the grain size of the zirconia-hafnia solid solution particles is too large, *i.e.* above the critical particle size for the spontaneous tetragonal to monoclinic transformation on cooling from the sintering temperature. Ideally, the dehydration of co-precipitated hydrous oxides is completed at temperatures as low as possible. Freeze drying is the best technique for demonstrating that ceramic powders dried at low temperatures exhibit a minimum degree of hard agglomeration and therefore a high degree of sinterability [12]. However, the low production efficiency of freeze drying largely limits its application on an industrial scale.

In comparison, conventional heating/drying of ceramic powders is simple and inexpensive. However, hard agglomerates may form in a conventionally dried ceramic powder. Microwave drying has the advantages of fast, high-energy efficiency, and produces a uniform residual moisture distribution in the dried powders. Organic solvent dehydration of hydrous oxides does not involve a significantly high temperature. It is therefore attractive in terms of avoiding the formation of any hard agglomerates in the dried powders. The aim of the present work is to investigate the impact of various drying routes on the characteristics of co-precipitated zirconia-hafnia powders.

2. EXPERIMENTAL WORK

The composition $Hf_{0.25}Zr_{0.75}O_2$ was chosen for this work because the critical particle size for spontaneous tetragonal to monoclinic transformation in this composition is in the range 50 to 100 nm. The sintering of a nanosized $Hf_{0.25}Zr_{0.75}O_2$ powder at temperatures below 1300°C may result in a grain size which is comparable to this critical particle size and therefore a dense ceramic body. Zirconium oxynitrate solution and hafnium oxynitrate powder, or hafnyl nitrate, were used as the starting materials. Table 1 lists the chemical analysis for the as-received zirconium oxynitrate solution and hafnium oxynitrate powder, respectively. Appropriate amounts of zirconium oxynitrate solution and hafnium oxynitrate powder were weighed out on the basis of the desired composition and mixed together in distilled water to obtain a 0·1 M

**Table 1. Chemical analyses of as-received zirconium
oxynitrate solution and hafnium oxynitrate powders
from Magnesium Elektron Ltd., Twickenham, U.K.
and Teledyne Wah Chang, Alabany, U.S.A.,
respectively**

	$ZrO(NO_2)_2xH_2O$	$HfO(NO_3)_2 2H_2O$
ZrO_2	19·6 wt%	0·91 wt%
HfO_2	0·4 wt%	53·5 wt%
SiO_2	< 100 ppm	< 50 ppm
Al	n/a	< 35 ppm
Cu	n/a	< 20 ppm
Fe	100 ppm	50 ppm
Na	< 100 ppm	n/a
Ti	n/a	< 25 ppm
Zn	n/a	< 50 ppm

salt solution. Co-precipitation was carried out by dropping the salt solution
from a burette pipette in to an aqueous ammonia solution of pH 10·5, while
the ammonia solution was vigorously stirred. The co-precipitated gel
suspension was aged for 24 hours. The hydroxide gels were then filtered out
from the aqueous suspension using Whatman 542 filter papers with the
assistance of a vacuum pump. The hydrous oxide cake was then repeatedly
washed and filtered six times, till the nitrate ions in washing solution reached
trace level.

The washed hydroxide gels were dried using three different routes including
organic solvent dehydration using acetone-toluene, microwave drying, and
conventional IR heating/drying. These three approaches are referred as OR,
MW, and IR, respectively, in the following discussions. For the OR drying, the
hydrous oxide cake was repeatedly acetone washed four times, followed by
toluene and then acetone washes, to dehydrate the hydroxide gels. The organic
solvent washed cake was then placed and kept in a vacuum flask for 24 hours
to evaporate off the organic solvents. In the MW route, the hydrous oxide
cake was dried in a commercially available Sharp 430 microwave oven which
generates 650 W power output at 2450 MHz at a 5·5 A input current, till the
weight loss in a 5 minute interval is negligible. For the conventional IR drying,
the hydrous oxide cakes was dried under an IR heating lamp, till the weight
loss is negligible.

The dried hydroxide powders were then calcined for 4 hours in air at various
temperatures in the range of 550°C and 1150°C using a heating rate of 3°C per
minute. Ball milling in ethanol for 4 hours using zirconia balls as milling
media was used to eliminate large agglomerates in the calcined powders, prior
to compaction uniaxially at 50 MPa and then isostatically at 200 MPa. The
powder compacts were finally fired at temperatures in the range of 1300°C and
1550°C for 2 hours using heating and cooling rates of 10°C per minute.

The characteristics of both the hydroxide and oxide powders in various
stages described above are monitored using XRD for phase identification,
BET for surface area, DTA and TGA for thermal behaviour, sedigraph, SEM,
and TEM for particle size and particle morphology.

3. RESULTS AND DISCUSSION

The as-precipitated hydroxide was amorphous, as was indicated by XRD phase analysis, Figure 1. A calcination at 600°C for 4 hours led to a significant phase change, in association with dehydration, nucleation and crystal growth of zirconia-hafnia oxide. As is shown in Table 2, there is a fall in BET specific surface area and a large rise in crystallite size when $Hf_{0.25}Zr_{0.75}O_2$ powder particles form at the calcination temperature, regardless of the drying routes employed. Monoclinic was the only phase present in the $Hf_{0.25}Zr_{0.75}O_2$ oxide on completion of the calcination at 600°C for 4 hours. This is in agreement with the experimental work performed by Tau and co-workers [13], who observed 100% monoclinic phase in the co-precipitated $Hf_{0.25}Zr_{0.75}O_2$ powder. These authors showed that up to 100% tetragonal phase could be retained in monolithic zirconia powders prepared using the identical precipitation route when the pH value of supernatant liquid was controlled in certain ranges. The results obtained by Tau et al. [13], together with the XRD phase analysis results obtained in the present work indicate the difficulty of retaining a metastable tetragonal phase in zirconia-hafnia solid solutions.

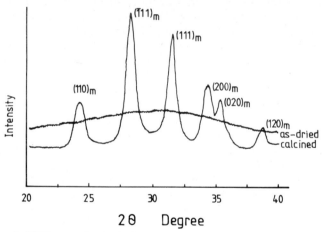

Figure 1. XRD traces for both the organic solvent dehydrated hydrous $Hf_{0.25}Zr_{0.75}O_2$ oxide and the oxide calcined at 600°C for 4 hours.

Table 2. BET specific surface areas for hydroxides and calcined oxides processed using organic solvent dehydration, microwave drying, and conventional heating drying. The crystallite size of oxide was worked out on the basis of BET surface area

Drying route	BET surface area (m²/g) Hydrous oxide	Oxide	Crystallite size of calcined oxide (nm)
OR	208·9	33·1	26·4
MW	112·4	26·4	33·0
IR	110·3	23·9	36·5

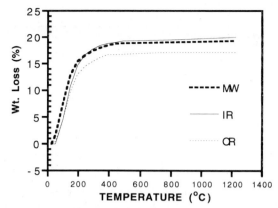

Figure 2. TGA traces for the hydrous oxides processed using organic solvent dehydration, microwave drying, and conventional heating drying.

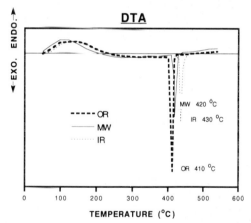

Figure 3. DTA traces for the same materials shown in Figure 2. The crystallization of $Hf_{0.25}Zr_{0.75}O_2$ oxide occurred immediately after the dehydration was completed.

It was demonstrated both by DTA and TGA studies that the thermal behaviour of hydrous oxides varies with different drying routes, as shown in Figures 2 and 3. Weight loss of 15 to 20 wt% occurred in the temperature range of 50 to 300°C for all hydrous oxides, regardless of the difference in drying method. The microwaved and conventional heating dried powders exhibit very similar thermal losses when heated at a rate of 10°C per minute. In comparison, the organic solvent dehydrated powder exhibits 3 to 4% less weight loss than either microwave dried or conventional heating/dried powders. These TGA results imply that organic solvent dehydration is a more effective dehydration approach than either microwave drying or conventional heating/drying. Figure 3 shows that crystallization of $Hf_{0.25}Zr_{0.75}O_2$ occurred almost immediately after the dehydration was completed. Drying route had

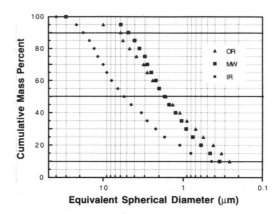

Figure 4. The particle size distributions measured using sedigraph in $Hf_{0.25}Zr_{0.75}O_2$ powders processed using various drying routes. The organic solvent dehydration dried powder exhibits the finest average particle size.

little impact on the crystallization temperature of dehydrated oxides. The crystallization temperature is 410°C, 420°C, and 430°C for OR, MW and IR dried powders, respectively.

As is shown in Table 2, the characteristics of both partly dehydrated hydrous oxide and calcined oxide are affected by the drying route employed to dehydrate the co-precipitated zirconium and hafnium hydrous gels. The organic solvent dehydration results in a finer crystallite size than either microwave drying or conventional heating/drying, in terms of BET surface area measurement and crystallite size determination.

The secondary particle size distribution in the calcined and then milled powders prepared using the three different drying routes is shown in Figure 4. These results may not truly represent the particle size distribution in each powder, as certain particle agglomerates might not be fully dispersed during the process of sample preparation for sedigraph measurement. However, they do indicate the average degree of particle agglomeration in these zirconia-hafnia powders. Therefore, the average particle size in each powder may be smaller than the particle size shown in Figure 4. The powder processed using the conventional heating drying route exhibits a larger average particle size and a wider particle size distribution than either the powder processed using the microwave drying route or the organic solvent dehydration route. The powder processed using the organic solvent dehydration route exhibits the finest average particle size. For example, the average particle size is 6 μm, 1·5 μm and 1·3 μm for the powders processed using IR, MW and OR drying routes, respectively.

Figure 5 and Table 3 show the tap and compacted green densities as a function of uniaxial compaction pressure for the powders dried using the three different approaches. The organic solvent dehydrated powder has a much lower tap density than either the microwave dried powder or the conventional heating/dried powder. The compacted density of the organic solvent

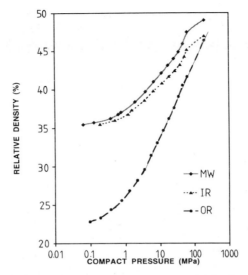

Figure 5. The green density as a function of compaction pressure for the powders processed using different drying routes.

Table 3. Tap density, green density, and agglomerate strength for the powders processed using various drying routes

Drying route	OR	MW	IR
TAP density (% theoretical density)	26·3	37·5	36·8
Green density (% theoretical density)	46·4	49·0	48·4
Agglomerate strength (MPa)		40	40

dehydrated powder is also much lower than that of either the microwave dried powder or the conventional heating/dried powder when the compaction pressure is below 100 MPa. With increasing compaction pressure over the pressure range 1 to 100 MPa, the compacted green density of the organic solvent dehydrated powder increases dramatically. At 100 MPa, the three powders exhibit similar compacted densities, although the microwave dried powder compact is slightly denser than the other two. It is known that the green density of a ceramic powder compact is largely dependent on the characteristics of the powder, such as particle size, particle size distribution, crystallite size, and the degree of powder agglomeration [14, 15]. For example, a highly crystallized ceramic powder exhibits a higher compacted density than a finely divided or an amorphous ceramic powder. The presence of hard agglomerates in a ceramic powder has a strong influence on the pressure dependence of compacted density. As discussed above, the average crystallite and particle sizes of organic solvent dehydrated powder are smaller than those of either microwave dried or conventional heating/dried powders. The tap and compacted densities of the organic solvent dehydrated powder are thus lower than those of the microwave dried or conventional heating/dried

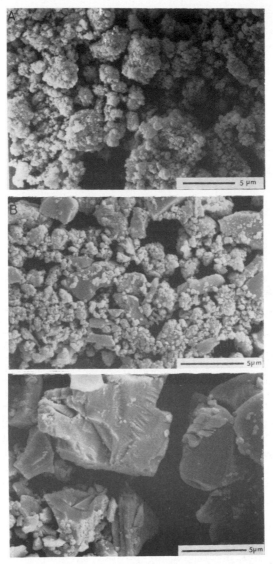

Figure 6(a, b, c). SEM micrographs showing the powders processed using organic solvent dehydration, microwave drying, and conventional heating drying, respectively.

powders. It is estimated from the pressure dependence of compacted density that the compaction strength of particle agglomerates is 40 MPa, for both the microwave dried and conventional heating/dried powders.

It was observed using SEM that the crystallite size and degree of particle agglomeration vary significantly in the powders processed using different drying routes. Figure 6(a, b, c) are three SEM micrographs showing the powders processed, using organic solvent dehydration, microwave drying and conventional heating/drying, respectively. The organic solvent dehydrated powder consists of particle agglomerates of 1 to 5 μm. The primary particle sizes are in the range of 0·1 to 0·3 μm. The microwave dried powder exhibits a duplex mode consisting of powder agglomerates of submicron sized particles and relatively large crystals of 1 to 4 μm. By contrast, the primary particle size of the conventional heating/dried powder is in the range of 3 to 8 μm. These powder particles are fractured from even larger particles when they were ball milled. The large oxide crystals exhibit a much more angular particle morphology, together with clearly visible fracture surfaces.

Figures 7(a, b, c) are three bright field TEM micrographs further showing the crystallite and primary particle sizes in the three different powders. Firstly, the organic solvent dehydrated powder consists of uniformly sized individual zirconia-hafnia crystals of 30 to 50 nm in size. Some of these microcrystals are well dispersed and the others from soft agglomerates. In the microwave dried powder, however, these microcrystals are strongly bridged together. The neck bonding is seen between individual crystals, implying that a diffusion was taking place when the powder was calcined at 600°C. The large $Hf_{0.25}Zr_{0.75}O_2$ crystals exhibit a twinning structure, indicating that they are monoclinic phase. In the conventional heating/dried powder, the nanocrystals are also strongly aggregated together by a bonding neck, forming large powder lumps.

A preliminary study has been made of the sinterability of these three powders by sintering the powder compacts at 1300°C for 2 hours. Table 4 shows the sintered density for the powders processed using organic solvent dehydration, microwave drying, and conventional heating drying, respectively. The conventional heating dried powder compact exhibited an unsintered powder-like structure on sintering at 1300°C for 2 hours. This indicates that little densification occurred in this material at 1300°C. Similarly, the microwave dried powder compact was not properly sintered, although its sintered density is slightly higher than that of the conventional heating dried powder compact. By comparison, the organic solvent dehydrated powder compact exhibited a sintered density of 84·8% theoretical density. The sintered sample was heavily cracked on cooling from the sintering temperature. It is estimated that the sintered density of this material is above 90% theoretical density provided it did not seriously crack on cooling from the sintering temperature. As was confirmed using XRD phase analysis, spontaneous tetragonal to monoclinic transformation had occurred on cooling from the sintering temperature in this sample. The volume expansion and shear deformation associated with the transformation results in the formation of a crack network.

Figure 8 is a plot showing the crystallite/particle size as a function of calcination temperature for $Hf_{0.25}Zr_{0.75}O_2$ processed using organic solvent dehydration. The crystallite/particle size increases steadily with increasing calcination temperature over the range of 600 to 1150°C. The particle

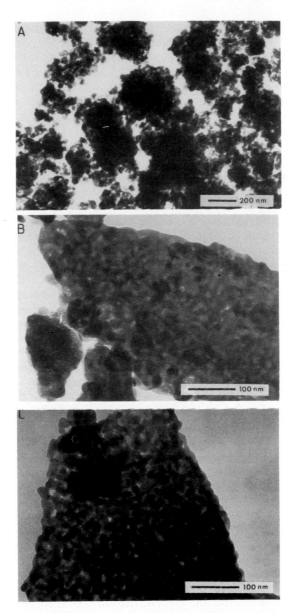

Figure 7(a, b, c). Bright TEM micrographs showing the powders processed using organic solvent dehydration, microwave drying, and conventional heating drying, respectively. The submicron crystals are loosely agglomerated together in the organic solvent dehydrated powder. However, the aggregation of submicron $Hf_{0.25}Zr_{0.75}O_2$ crystals is observed both in the microwave dried and in the conventional heating dried powders.

Table 4. The sintered density at 1300°C for 2 hours for the powders processed using different drying routes

Drying route	Sintered density (% theoreticald density)
OR	84·8
MW	63·4
IR	56·6

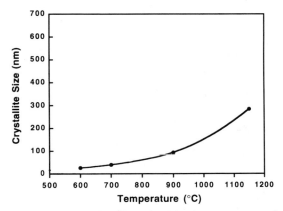

Figure 8. The crystallite size as a function of calcination temperature for the organic solvent dehydrated powder.

coarsening is a direct result of crystal/particle aggregation at the calcination temperature. It was observed using SEM and TEM that the neck thickness between submicron $Hf_{0.25}Zr_{0.75}O_2$ crystals increased dramatically with rising calcination temperature.

4. CONCLUSIONS

Co-precipitation of mixed zirconium and hafnium oxynitrates in an ammonia solution at pH 10·5 results in the formation of amorphous zirconium-hafnium hydroxide. The hydrous oxide changed to monoclinic $Hf_{0.25}Zr_{0.75}O_2$, on calcining at 600°C for 4 hours. However, the characteristics of the calcined powder, such as crystallite size, particle size, particle size distribution, and the degree of powder agglomeration, are influenced strongly by the drying route which is employed to dehydrate the co-precipitated hydrous gels. The organic solvent dehydrated powder exhibits smaller average crystallite and particle sizes, and a lower degree of powder agglomeration than the powders processed using either microwave drying or conventional heating/drying routes. The fine particle/crystal size and a low degree of powder agglomeration lead to a high sinterability and a low sintering temperature.

ACKNOWLEDGMENTS

J. Wang wishes to thank the SERC for finanical support during the period when this work was done.

REFERENCES

1. GARVIE, R. C., HANNINK, R. H. J. & PASCOE, R. T., *Nature,* **258**, 703, (1975).
2. CLAUSSEN, N., *Mater. Sci. Eng.,* **71**, 23, (1985).
3. CLAUSSEN, N., RUHLE, M. & HEUER, A. H., *Advances in Ceramics, Vol. 12, "Science and Technology of Zirconia II,"* publ. The American Ceramic Society, Columbus, OH, (1984).
4. WANG, J., LI, H. P. & STEVENS, R., *Hafnia and Hafnia Toughened Ceramics,* accepted for publication in *Journal of Materials Science,* (1992).
5. TIEN, T. Y., BROG, T. K. & LI, A. K., *Intl. J. High Tech. Ceram.,* **2**, 207, (1986).
6. HUNTER, Jr., O., SCHEIDECKER, R. W. & TOJO, S., *Ceram. Intl.,* **5**, 137, (1979).
7. BAILEY, J. E., LEWIS, D., LIBRANT, Z. M. & PORTER, L. J., *J. Brit. Ceram. Soc.,* **71**, 25, (1972).
8. RUH, R., GARRETT, H. J., DOMAGALA, R. F. & TALLAN, N. M., *J. Amer. Ceram. Soc.,* **51**, 23, (1968).
9. KRIECHBAUM, G. W., KLEINSCHMIT, P. & PEUCKERT, D., *Ceram. Trans.,* **1**, 126, (1988).
10. DOLE, S. L., SCHEIDECKER, R. W., SHIERS, L. E., BERARD, M. F. & HUNTER, Jr., O., *Mater. Sci. Eng.,* **32**, 277, (1978).
11. RHODES, W. H., *J. Amer. Ceram. Soc.,* **64**, 19, (1981).
12. JOHNSON, D. W. & SCHNETTLER, F. J., *J. Amer. Ceram. Soc.,* **53**, 440, (1970).
13. TAU, L. M., SRINIVASAN, R., DE ANGELIS, R. J., PINDER, T. & DAVIS, B. H., *J. Mater. Res.,* **3**, 561, (1988).
14. KENDALL, K., *Powder Metall.,* **31**, 28, (1988).
15. REED, J. S. & RUNK, R. B., in *Treatise on Materials Science and Technology, 9: Ceramic Fabrication Processes,* Ed. F. F. Y. Wang, Academic Press, NY, (1976), p. 71.

Nitrogen Stabilization of Tetragonal Zirconias

B. A. SHAW, Y. CHENG and D. P. THOMPSON

Wolfson Laboratory, Materials Division, Department of Mechanical, Materials and Manufacturing Engineering, University of Newcastle upon Tyne, NE1 7RU, UK

ABSTRACT

Previous work at Newcastle has shown that a 3 $^m/o$ yttria-stablized zirconia reacts with nitrogen when heated at temperatures above 1400°C to form a non-transformable tetragonal (t') product due to partial substitution of oxygen by nitrogen. The present paper represents a continuation of this work, and describes the incorporation of nitrogen into zirconia containing a wide range (0·25–8·0 $^m/o$) of yttria additions. Product phases have been characterized by X-ray diffraction and show a consistent trend with variable yttria content. The results substantiate the explanation given previously [1, 2] for the poor toughening performance of zirconia in nitrogen ceramic matrices, compared with its behaviour in oxide systems.

1. INTRODUCTION

When pure zirconia or a mixture of zirconia and ceramic nitrides are sintered in a nitrogen atmosphere at > 1400°C it has been observed that partial substitution of nitrogen for oxygen can occur. Early work by Gilles *et al.* [3, 4] reported that the solid-state reaction of zirconia with zirconium nitride produced three different ordered zirconium oxynitride phases (β', β and γ) which correspond to different superstructures of the basic zirconia fluorite-type lattice; the β' and β phases have a rhombohedral structure and the γ phase has a cubic structure (see Figure 1). More recent work by Claussen *et al.* [5] proposed that the cubic form of zirconia could be stabilized by nitrogen, which was incorporated when reacting monoclinic zirconia with various metallic nitrides in a nitrogen atmosphere. It is more likely, however, based on more recent work, that what was actually observed on X-ray patterns of the product, were the strong lines of the β' zirconium oxynitride, which correspond exactly to those of cubic zirconia.

Over the past few years, work at Newcastle has shown that nitrogen can be incorporated into the transformable tetragonal structure, t, of a 3 $^m/o$ yttria-stabilized zirconia leading to further stabilization to give the non-transformable tetragonal, t', and eventually a cubic, c, structure [1, 2] (see Figure 2 — in this and subsequent figures, * indicates a tetragonal phase, ■ a cubic phase and □ a cubic phase where tetragonal was also observed). It has also been found that solid-state reaction (*e.g.* ZrO_2 + ZrN) gives an easier route for nitrogen incorporation than molecular nitrogen alone [6]. From this work it appears that the pre-existing disordered vacancies in the zirconia structure, introduced by the Y^{3+} cation, tend to suppress the formation of ordered zirconium oxynitride phases. This is probably because of the difficulty in re-ordering these vacancies.

Nitrogen incorporation into zirconia, resulting in a non-transformable product, also occurs in more complex systems such as zirconia-toughened silicon nitride or zirconia-toughened sialons and this may be a fundamental

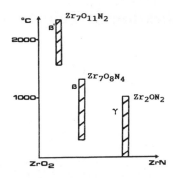

Figure 1. Composition and stability of the ordered Zr-O-N phases [3].

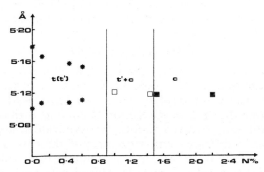

Figure 2. Variation of lattice parameters with nitrogen content for a 3 ^{m}/o Y_2O_3 stabilized zirconia.

reason why zirconia additions to nitrogen containing ceramics do not result in the significant increases in toughness observed in oxide ceramics (see Figures 3 and 4).

The aim of the present work was to try and gain a better understanding of the Zr-Y-O-N system by repeating the same type of experiments as those previously carried out on the 3 ^{m}/o yttria-stabilized zirconia, using powders stabilized with different amounts of yttria.

2. EXPERIMENTAL

The powders used in these experiments ranged from pure, monoclinic, m, zirconia up to an 8 ^{m}/o yttria-stabilized (cubic) zirconia. Commercial zirconia powders containing 0, 2, 4, 6, and 8 ^{m}/o yttria were used (Tosoh Corporation, Tokyo, Japan — grades TZ-2Y, TZ-4Y, TZ-6Y and TZ-8Y) as well as zirconia powders containing smaller yttria additions of 0·25, 0·50, 0·75 and 1 ^{m}/o which were prepared using a sol-gel technique.

The sol-gel technique used was the chloride synthesis process which is described in detail by Readey *et al.* [9]. Basically the technique involves co-precipitating a gel of $Zr(OH)_4$ and $Y(OH)_3$ by adding a chloride solution dropwise into a stirred NH_4OH solution (pH > 10) (see Figure 5), the final powder being produced from the gel by washing with deionised water followed

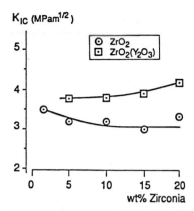

Figure 3. Fracture toughness as a function of the volume fraction of ZrO₂ in a Si₃N₄ matrix [7].

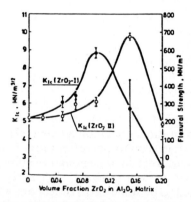

Figure 4. Fracture toughness as a function of the volume fraction of ZrO₂ in an Al₂O₃ matrix [8].

by calcination and milling. The powders produced for these experiments were given a final wash with ethanol prior to calcination in order to prevent the formation of hard agglomerates [10].

In the nitriding experiments, cold pressed pellets of the various zirconia powders (weighing about 1 g) were placed in boron nitride lined graphite crucibles and sintered, in a graphite-element resistance furnace, using a 30°C/min heating rate and natural furnace cooling. Each sample was sintered for 2 hrs, in a flowing nitrogen atmosphere, at temperatures ranging from 1600°C to 2000°C.

After sintering, the weight change of each sample was measured and then each sample was crushed into a powder to allow X-ray powder photographs to be taken using a Hägg-Guinier X-ray focusing camera and monochromatic CuKα₁ radiation. A photograph of the raw powder, after calcination at 1100°C for 4 hrs, was also taken. From the X-ray photographs the cell dimensions of the resulting zirconia phases were measured and the

Table 1. Nitriding results for pure ZrO₂

Sample	$\Delta W(\%)$	$N(^w/o)$	Phases
Raw powder	—	—	m(vs)
1600°C	−0·82	0·09	m(vs), $\beta''(\beta')(vw)$
1700°C	−0·94	0·16	m(vs), $\beta''(\beta')(w)$
1800°C	−1·02	0·40	m(vs), $\beta''(\beta')(m)$
1900°C	−1·65	1·08	m(s), $\beta''(\beta')(ms)$
2000°C	−2·95	2·24	m(ms), $\beta''(\beta')(s)$, ZrN(w)

transformability of any tetragonal phase present was assessed. Finally, an analysis of the nitrogen content in each sample was carried out by gas chromatography, using a Carlo Erba combustion analyser.

3. RESULTS AND DISCUSSION

3.1 Nitriding of Pure Zirconia

Table 1 shows that, consistent with the work carried out by Gilles *et al.* [3, 4], when pure zirconia is nitrided, the nitrogen incorporated into the structure does not give stabilization, but instead forms an ordered zirconium oxynitride phase. The type of zirconium oxynitride phase formed was characterized by examining the position of the weak superlattice lines on the X-ray photographs. Unfortunately, below 1800°C, the small amount of oxynitride meant that these lines were too weak to be seen and only the stronger fluorite-type reflections indicated that an oxynitride phase was present. The large number of X-ray lines from the monoclinic zirconia also complicated matters by overlapping with possible oxynitride reflections. However, at and above 1800°C the superlattice lines were strong enough to allow the oxynitride phase to be identified. As shown, as well as the β′ phase (reported by Gilles *et al.* [3, 4]) there also appeared to be some extra reflections which corresponded to another oxynitride phase, β″, observed in previous work at Newcastle [1].

At present, little is known about these zirconium oxynitride phases, although, from the results obtained it seems that the β″ phase contains less nitrogen than the β′ phase. It is known, however, that these phases have poor oxidation resistance [11, 12] and that they are non-transformable to the monoclinic form and therefore, they must be avoided if zirconia-toughened nitrogen ceramics are to be produced.

Table 2. Nitriding results for pure ZrO₂ + 0·50 ᵐ/o Y₂O₃

Sample	$\Delta W(\%)$	$N(^w/o)$	Phases
Raw powder	—	—	m(vs)
1600°C	−3·19	0·44	m(vs), $\beta''(\beta')(w)$
1700°C	−2·74	0·53	m(vs), $\beta''(\beta')(m)$
1800°C	−3·27	1·02	m(s), $\beta''(\beta')(ms)$
1900°C	−3·71	2·10	m(ms), $\beta'(s)$
2000°C	−5·64	2·28	m(m), $\beta'(vs)$, ZrN(m)

3.2 Nitriding of Yttria-Stabilized Zirconias Containing $\leqslant 1$ ^{m}/o Y_2O_3

The results obtained for the 0·25, 0·50, 0·75 and 1 ^{m}/o yttria-stabilized zirconia samples were all very similar to those for the pure zirconia in that the nitrogen incorporated into the structure formed an ordered zirconium oxynitride phase rather than giving t or t′ products. Table 2 gives the results for the 0·50 ^{m}/o yttria samples (the 0·25, 0·75 and 1 ^{m}/o samples were all very similar). It is interesting to note that, at a given temperature, the amount of nitrogen incorporated into the yttria-containing samples was greater than that incorporated into the pure zirconia samples. This was most probably due to the pre-existing vacancies in the anion lattice of the yttria-containing zirconias, facilitating an easier and quicker diffusion path for the nitrogen.

3.3 Nitriding of Yttria-Stabilized Zirconias Containing $\geqslant 2$ ^{m}/o Y_2O_3

Tables 3–6 and Figures 6–9 show the results obtained when the 2, 4, 6 and 8 ^{m}/o yttria-stabilized zirconia samples were nitrided. From these results it can be seen that in each series of samples, nitrogen incorporation did not lead to the formation of any ordered zirconium oxynitride phases. As stated earlier, the reason for this can be attributed to the difficulty in re-ordering the pre-existing anion vacancies (introduced into the zirconia lattice when the Y^{3+} cations are incorporated) which is necessary for the ordered oxynitride phases to form. The results show that somewhere between a 1 and 2 ^{m}/o yttria addition to the zirconia is sufficient to suppress the formation of the ordered oxynitride phases.

From a consideration of the low-yttria end of the ZrO_2-Y_2O_3 phase diagram (see Figure 10) and by examining the phases that were formed at each

Table 3. Nitriding results for ZrO_2 + 2 ^{m}/o Y_2O_3

Sample	$\Delta W(\%)$	$N(^W/o)$	Phases	$a_c(\text{Å})$	$a_t(\text{Å})$	$c_t(\text{Å})$	c_t/a_t
Raw power	—	—	m(w), t(vs)	—	5·098	5·180	1·016
1600°C	−1·19	0·12	m(ms), t(s)	—	5·100	5·181	1·016
1700°C	−1·27	0·27	m(m), t(s)	—	5·101	5·179	1·015
1800°C	−1·46	0·40	m(mw), t(vs)	—	5·101	5·167	1·013
1900°C	−2·07	1·01	m(vw), t(s), c(m)	5·112	5·100	5·171	1·014
2000°C	−3·72	2·38	t(w), c(vs), ZrN(w)	5·113	5·108	5·171	1·012

Table 4. Nitriding results for ZrO_2 + 4 ^{m}/o Y_2O_3

Sample	$\Delta W(\%)$	$N(^W/o)$	Phases	$a_c(\text{Å})$		$a_t(\text{Å})$	$c_t(\text{Å})$	c_t/a_t
Raw power	—	—	t(vs)	—		5·109	5·172	1·016
1600°C	−1·04	0·16	m(m), t_1(ms), t_2(m)	—	(1)	5·120	5·157	1·007
					(2)	5·098	5·182	1·016
1700°C	−1·08	0·20	m(mw), t_1(s), t_2(w)	—	(1)	5·115	5·157	1·008
					(2)	5·097	5·177	1·016
1800°C	−1·18	0·41	m(vw), t_1(vs)	—		5·114	5·149	1·007
1900°C	−1·81	1·11	c(vs), t(vw)	5·122		—	—	—
2000°C	−3·23	2·32	c(s), ZrN(m)	5·121		—	—	—

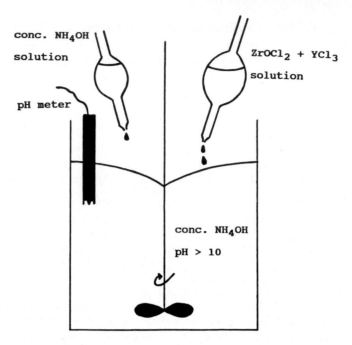

Figure 5. Method used to co-precipitate a gel of $Zr(OH)_4$ + $Y(OH)_3$ from a chloride solution.

Table 5. Nitriding results for ZrO_2 + 6 ^{m}/o Y_2O_3

Sample	$\Delta W(\%)$	$N(^w/o)$	Phases	$a_c(\text{Å})$	$a_t(\text{Å})$	$c_t(\text{Å})$	c_t/a_t
Raw power	—	—	c(s), t(ms)	5·135	5·121	5·165	1·009
1600°C	−0·94	0·16	t(vs)	—	5·125	5·155	1·006
1700°C	−1·00	0·26	t(vs)	—	5·124	5·154	1·006
1800°C	−1·16	0·39	c(s), t(w)	5·132	5·126	5·149	1·004
1900°C	−2·28	1·14	c(vs)	5·131	—	—	—
2000°C	−3·57	1·96	c(vs), ZrN(mw)	5·128	—	—	—

Table 6. Nitriding results for pure ZrO_2 + 8 ^{m}/o Y_2O_3

Sample	$\Delta W(\%)$	$N(^w/o)$	Phases	$a_c(\text{Å})$
Raw powder	—	—	c(vs)	5·142
1400°C	−1·60	0·09	c(vs)	5·139
1500°C	−1·00	0·14	c(vs)	5·139
1600°C	—	0·19	c(vs)	5·139
1700°C	−1·10	0·29	c(vs)	5·137
1800°C	−1·40	0·80	c(vs)	5·135
2000°C	−3·50	2·23	c(vs), ZrN(ms)	5·133

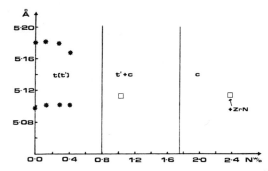

Figure 6. Variation of lattice parameters with nitrogen content for a 2 ^{m}/o Y$_2$O$_3$ stabilized zirconia.

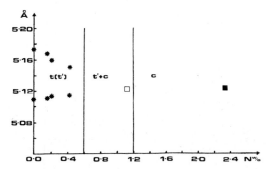

Figure 7. Variation of lattice parameters with nitrogen content for a 4 ^{m}/o Y$_2$O$_3$ stabilized zirconia.

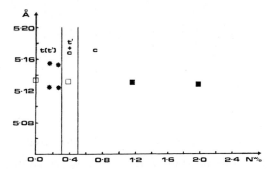

Figure 8. Variation of lattice parameters with nitrogen content for a 6 ^{m}/o Y$_2$O$_3$ stabilized zirconia.

temperature for each set of results, it can be seen that as the temperature is increased (giving an increased driving force for nitrogen incorporation) there is a general trend for the zirconia to become increasingly stabilized. As the temperature was increased in the 2 and the 4 ^{m}/o yttria-containing samples, the t' content increased while the amount of t (shown as monoclinic after grinding) decreased and eventually the cubic phase started to form. Clearly the

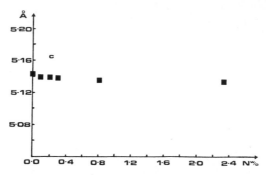

Figure 9. Variation of lattice parameters with nitrogen content for an 8 ^{m}/o Y_2O_3 stabilized zirconia.

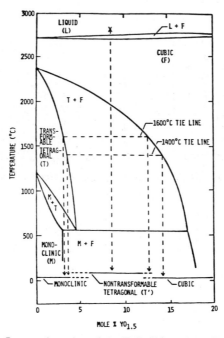

Figure 10. Low-yttria region of the ZrO_2-$YO_{1.5}$ phase diagram [13, 14].

nitrogen is giving additional stabilization because the cubic phase would not have been expected to form at all in these samples. The formation of two different tetragonal phases, at the lower temperatures, in the 4 ^{m}/o yttria-stabilized zirconia, were probably caused by a non-equilibrium situation where a gradient of nitrogen had formed across the zirconia grains due to an insufficient driving force for the nitrogen diffusion. This is consistent with previous work [2]. The stabilization process is also well illustrated in the results for the zirconia samples containing 6 ^{m}/o yttria where, with increasing

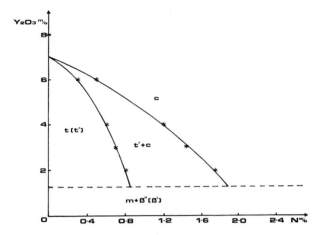

Figure 11. Phases observed in the Zr-Y-O-N system as a function of Y_2O_3 and N content (after fast cooling from $> 1600°C$).

Table 7. Vacancies required to stabilize the cubic form of zirconia using Y^{3+} and N^{3-} as stabilizing ions

$Y_2O_3(^m/o)$	Approx. $N(^w/o)$	Calculated formula*	% vacancies in the anion lattice
2	1·75	$Zr_{0.961}Y_{0.039}O_{1.752}N_{0.152} \square_{0.096}$	4·80
3	1·5	$Zr_{0.942}Y_{0.058}O_{1.776}N_{0.130} \square_{0.094}$	4·70
4	1·2	$Zr_{0.926}Y_{0.077}O_{1.806}N_{0.104} \square_{0.090}$	4·50
6	0·5	$Zr_{0.906}Y_{0.113}O_{1.877}N_{0.044} \square_{0.079}$	3·95
8	0	$Zr_{0.852}Y_{0.148}O_{1.926} \square_{0.074}$	3·70

*Anion vacancies represented by $\square$

temperature, the structure was seen to change from t' to $t' + c$ to c; however, without nitrogen present a t' structure would have been expected to form at all of the temperatures studied. A cubic structure was always observed in the $8 ^m/o$ yttria-stabilized samples. However, nitrogen incorporation still took place and this was shown by a small decrease in the cubic lattice parameter. In these samples the amount of nitrogen incorporated was quite low. This may be explained by the large number of vacancies, already existing in the zirconia, providing a resistance to the introduction of more vacancies by nitrogen incorporation. Clearly there must be a limit to the number of vacancies that can exist in the zirconia anion lattice and when this limit is reached then zirconium nitride should start to form.

The results presented show that nitrogen can enter either the tetragonal or cubic structure of a yttria stabilized zirconia to cause further stabilization, as shown in Figure 11. The "nitrogen stabilization" process is thought to be very similar to that by yttrium and this suggests that it is the number and distribution of oxygen vacancies that plays one of the key roles in determining the crystal structure of the zirconia phase formed. Table 7 shows that the cubic structure will form if between about 3·5 and 5·0% of the sites in the zirconia anion lattice are vacant. The exact value depends on how the stabilization is proportioned between the yttrium and the nitrogen; less vacancies are required if the yttrium content is high, *i.e.* yttria is a slightly stronger stabilizer than nitrogen. As stated previously, this can explain why, contrary to oxide ceramics, zirconia-toughening of nitrogen ceramics has not been observed.

4. CONCLUSIONS

1. Additions of between 1 and 2 ^{m}/o of yttria to zirconia are sufficient to make the ordering of anion lattice vacancies unfavourable and, hence, in nitriding conditions, ordered zirconium oxynitride phases do not occur.
2. Nitrogen can be incorporated into both the tetragonal and the cubic zirconia structures if at least $\approx$ 2 ^{m}/o of yttria is present. When this occurs, the N^{3-} anions give additional stabilization, in a similar way to the classical cation stabilizers, by introducing additional disordered vacancies into the anion sub-lattice.
3. The number and distribution of vacancies in the zirconia anion lattice is one of the main deciding factors controlling which structure the zirconia will assume and with between 3·5–5·0% of the oxygen sites vacant (dependent on the ratio of $Y^{3+}:N^{3-}$) the structure will become fully cubic.

ACKNOWLEDGMENTS

One of us (B.A.S.) gratefully acknowledges SERC funding under the CASE scheme in collaboration with the Cookson Technology Centre, Oxford, U.K. We would also like to thank Dr. J. Bultitude (Cookson Technology Centre) for his advice and help with the sol-gel process and Mr. D. Dunbar (Department of Chemistry, University of Newcastle upon Tyne) for carrying out the nitrogen analyses.

REFERENCES

1. CHENG, Y. & THOMPSON, D. P., *Special Ceramics, 9*, 149, (1992).
2. CHENG, Y. & THOMPSON, D. P., *J. Am. Ceram. Soc., 74*, 1135, (1991).
3. GILLES, J. C., *Bull. Soc. Chim. Fr., 22*, 2118, (1962).
4. COLLONGUES, R., GILLES, J. C., LEJUS, A. M., PEREZ, M., JORBA, Y. & MICHEL, D., *Mater. Res. Bull., 2*, 837, (1967).
5. CLAUSSEN, N., WAGNER, R., GAUCKLER, L. J. & PETZOW, G., *J. Am. Ceram. Soc., 61*, 369, (1978).
6. CHENG, Y., Sixth Progress Report, The University of Newcastle upon Tyne, October, (1991).
7. EKSTROM, T., FALK, L. K. L. & KNUTSON-WEDEL, E. M., *J. Mat. Sci., 26*, 4331, (1991).
8. CLAUSSEN, N., *J. Am. Ceram. Soc., 59*, 49, (1976).
9. READEY, M. J., LEE, R. R., HALLORAN, J. W. & HEUER, A. H., *J. Am. Ceram. Soc., 73*, 1499, (1990).
10. KALISZEWSKI, M. S. & HEUER, A. H., *J. Am. Ceram. Soc., 73*, 1504, (1990).
11. BABINI, G. N., BELLOSI, A., VINCENZINI, P., DALLE FABRICHE, D. & VISANI, R., *Science of Ceramics*, Ed. P. Vincenzini, publ. Ceramurgica, *12*, 471, (1984).
12. VINCENZINI, P., BELLOSI, A. & BABINI, G. N., *Ceram. Int., 12*, 133, (1986).
13. MILLER, R. A., SMIALEK, J. L. & GARLICK, R. G., Science and Technology of Zirconia, Eds. A. H. Heuer and L. W. Hobbs, publ. Am. Ceram. Soc., *Advances in Ceramics, 3*, 241, (1981).
14. SCOTT, H. G., *J. Mat. Sci., 10*, 1527, (1975).

Properties of Electrofused Yttria TZP Powders

H. A. J. THOMAS, K. SALMON and J. B. BUTTERS

Unitec Ceramics Ltd., Doxey Road, Stafford, ST16 1EA

ABSTRACT

This paper deals with some typical processing and mechanical properties demonstrated by Y-TZP ceramics produced from powders prepared by electrofusion, comminution and classification. The powder composition was 5 wt%, with a total impurity content of < 0·4 wt%, and the median particle size was measured as 0·57 µm.

Sintered densities, for uniaxial die pressed and isopressed samples, of the order of 99% theoretical were readily obtained at 1550°C. Three-point bend strength, toughness and hardness values could all be considered typical for such TZP materials. The results reported clearly demonstrate the competitive performance of powders prepared using electrofusion technology.

Reference is also made to the clear processing advantages to be found when using these powders in dispersion or plastic forming techniques.

1. INTRODUCTION

The realisation that the tetragonal to monoclinic phase change of zirconia could be exploited to produce tough materials, represented a significant advance in ceramic technology [1]. Considerable effort has since been dedicated towards developing material systems to derive the maximum benefit from the crack shielding characteristics of the transformation mechanism. A major consequence of this work was a family of materials commonly referred to as TZP (tetragonal zirconia polycrystals) ceramics [2, 3]. Such materials use a stabilising agent (typically yttria or ceria) to produce a fine, dense microcrystalline structure consisting almost entirely of metastable tetragonal phase. This maximises the transformation toughening process resulting in excellent room temperature strength, toughness and hardness.

Many approaches to the production of TZP powders have been attempted [4, 5]. The procedure which has demonstrated the most commercial success to date involves the coprecipitation of chemical precursors. However, the yttria stabilised TZP zirconia powder which is the subject of this study is produced by a commercially developed electrofusion and comminution method.

Production of zirconia powders by electrofusion is not a particularly new concept. Such materials, containing stabilising agents, have been available for many years. Until recently, however, these have only been considered suitable for refractory type applications. Continued development has resulted in considerable improvements in process control, to a level whereby sub-micron powders of high purity and homogeneity are possible. This has been applied, on a commercial scale, to zirconia powders containing calcia, magnesia, yttria and ceria as stabilising agents, since 1988. Properties of such powders have been dealt with in previous published work [6–10]. The present publication deals more specifically with a 5 wt% (nominally 3 mol%) yttria powder,

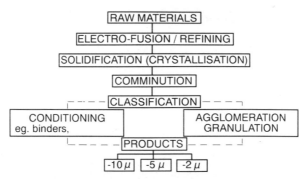

Figure 1. Process flow chart.

demonstrating a marked improvement in purity over these previous materials. It illustrates the competitive performance of these powders and highlights the perceived processing benefits.

2. EXPERIMENTAL DETAILS

Figure 1 outlines the route used to produce the powders for this study. High purity raw materials are blended to the predetermined composition and then electrofused using graphite electrodes. Control of the fusion process is critical for raw material refinement and production of an homogeneous finished product. The cooled ingot is broken down using controlled, contamination-free comminution and classification technology. The powders are finally dried and calcined ready for use. Particle size and surface area measurements were made with a Sedigraph 5100 and by single point BET method, respectively. Chemical analysis was carried out by means of X-ray fluorescence (XRF). Samples were prepared for testing by low pressure uniaxial die pressing at 8 MPa followed by isostatic pressing at 100 MPa. Sintering was achieved in a fibre lined kiln fitted with silicon carbide elements. Green densities were calculated from the dimensions and weight, whereas sintered densities were measured by the Archimedes principle, by suspension in water. Strength determinations were made using the Japanese 3-point bend testing standard JIS R1601 [11]. The bars were cut from large sintered discs which had been isopressed at 200 MPa.

Hardness was measured using the Vickers micro-indentation technique and the resultant cracks were used to calculate the fracture toughness by direct crack measurement. A 10 kg applied load was required to obtain cracks of a suitable length without spalling of the indent.

Scanning and transmission electron microscopy was used for the examination of powder morphology. Whereas phase analysis was carried out on powders and sintered surfaces by means of XRD.

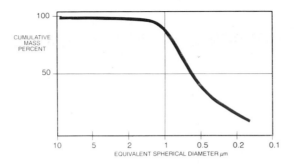

Figure 2. Particle size distribution of electrofused TZP fine powder.

3. RESULTS AND DISCUSSION

3.1 Powder characteristics

A typical particle size distribution curve of the finished powder is illustrated in Figure 2. It shows that 95% of the powder is less than 1·3 μm and it has a typical median size of 0·57 μm. The surface area of such a powder was found to be 6·4 m²/g.

The powder morphology, Figure 3, appears to show a typical submicron structure, of agglomerated fine particles. However, closer examination using TEM, reveals the high density of the individual powder grains, Figure 4. This feature is an intrinsic characteristic of all fused powders, and is fundamental to the resultant properties.

Figure 3. Typical morphology of a sub-micron electrofused powder.

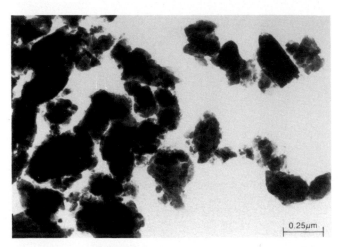

Figure 4. TEM micrograph showing the high density of the individual particles.

Table 1 compares a typical XRF chemical analysis for the current powder with that of an earlier grade [6]. It gives clear indication of the improved chemistry of the powders referred to in this paper. Attention is drawn, in particular, to the considerable reduction in the levels of SiO_2, Al_2O_3 and Fe_2O_3. This has been achieved by improved raw material selection and greater control of the fusion conditions.

The compaction behaviour of the raw powders, *i.e.* not spray dried, is illustrated in Figure 5, which shows the effect of pressure on density. The high green densities achieved are a further indication of the high grain density of the powder. Table 2 summarises other typical properties obtained with non-spray dried electrofused powders. The poor flow properties, emphasised by the low fill density and high compaction ratio, are typical of ultrafine powders. When pressing operations are involved it is advisable to use a spray dried derivative with improved flow characteristics.

Table 1. Chemical analysis of electrofused TZP powder

	Current high purity	*Standard purity*
$ZrO_2 + HfO_2$	94·52	93·51
Y_2O_3	5·21	5·75
SiO_2	< 0·02*	0·16
Al_2O_3	0·02	0·13
Fe_2O_3	0·02	0·09
TiO_2	0·15	0·17
Na_2O	< 0·1*	< 0·1*
CaO	0·01	0·04
MgO	< 0·05*	< 0·05*

*Limit of detection

**Table 2. Typical properties of
electrofused powders**

Median particle size (μm)	0·57
Specific surface area (m²/g)	6·4
Powder density (gcm⁻³)	5·82
Bulk density (gcm⁻³)	1·05
Tap density (gcm⁻³)	1·35
Compaction ratio (@ 100 MPa pressure)	2·86
% monoclinic in powder	30–40

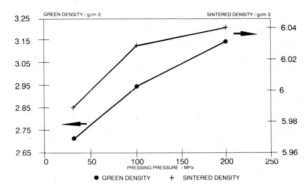

Figure 5. Density as a function of pressing pressure.

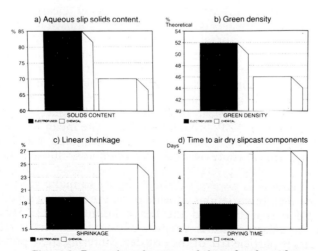

Figure 6. Processing advantages of electrofused powders.

Table 3. Properties of electrofused yttria TZP powders
sintered to 1550°C/2 hours at 5°C/min

Density (gcm^{-3})	6·02
Linear shrinkage (%)	20·2
Mean grain size (μm)	0·69
Bend strength (MPa)	912·8 ± 67·5†
Indentation toughness (MPam$^{1/2}$)	8·2
Vickers hardness, HV$_{10}$, (GPa)	11·8

†95% confidence limits

Figure 6 illustrates some of the clear processing advantages offered by an electrofused powder, especially when the material is to be used in slurries or formulations for plastic forming processes. The higher solids loading achievable, Figure 6(a), is reflected in the faster drying time of cast components, Figure 6(d). Resultant lower drying shrinkages leads to decreased drying strains, in conjunction with a reduced possibility of distortion and cracking. A further consequence of the high grain density is the lower requirement for organic additives. Thus, speeding up burn-out schedules and reducing factory emissions. As has already been mentioned previously, high grain density leads to high green density, Figure 6(b). The resultant lower shrinkage, Figure 6(c), should reduce the occurrence of distortion during firing. The combination of these factors can result in a significant reduction in processing time and reject losses.

3.2 Sintered properties

The green compacts were fired to 1550°C for 2 hours at a rate of 5°C/min. The sintered densities are also shown in Figure 5, as a function of the pressing pressure. Values in excess of 99% of theoretical clearly show that the sinteractivity of these powders is not severely compromised by the high particle density and lower surface area. Furthermore, the improved purity does not seem to have had a major effect on sinterability. Linear shrinkages, following pressing at 100 MPa, are of the order of 20–21%. It is worth mentioning, that the similarity of these shrinkages to those of technical aluminas should allow for a single set of tooling to be used for both materials.

Table 3 lists the mechanical properties typically observed for sintered samples of electrofused TZP zirconia powders. The mean bend strength and hardness are slightly lower than values usually quoted for TZP zirconia, derived from a chemical route, whereas the toughness value can be considered equivalent.

Strength in bending is always a difficult figure to make valid comparisons on, due to the inherent variations dependent on the sintering schedule used, as well as sample preparation and the test conditions [12]. However, it is possible that the observed lower strength of fused powders is related to the intrinsically larger particle size of the powder. Sintering these powders to high density could lead to grain sizes in excess of the critical size for restraining the transformation [3]. The resultant formation of monoclinic phase producing inherent microstructural weaknesses. The mean linear intercept grain size, at 0·69 μm, provides some evidence to support this premise, as it is significantly larger than that usually reported for TZP materials [3]. Though no direct evidence of transformed grains could be found on sintered or polished surfaces. XRD examination of as-sintered surfaces revealed the presence of 5% monoclinic phase, which was reduced to 2·6%, after diamond polishing to a 6 μm finish. These figures are higher than those usually quoted for a TZP microstructure, and may indicate an increased transformability.

However, this premise did not seem to be reflected in the toughness measurement. The value for DCM (Direct Crack Measurement) fracture toughness in Table 3 could be considered typical for a 3 mol% TZP powder. However, a great deal of variation can be observed in values measured by this technique due to the extreme sensitivity to sample preparation and test conditions. For comparison, therefore, a chemically derived powder was tested under the same conditions. The value obtained was 4 MP am$^{1/2}$, following a sintering schedule recommended in product literature.

The hardness value, in Table 3, is lower than values quoted in the literature for chemically derived powders. This may be a further indication of increased transformability, as hardness and strength seem to be inversely related to toughness. Although other controlling factors, such as density, may explain a low hardness measurement. It is interesting to note that a fused material with a sintered density of only 5·89 g cm^{-3} produced a mean strength of 955 MPa and had measured toughness and hardness values of 6·5 MP am$^{1/2}$ and 12·4 GPa respectively.

In summary, the values in Table 3 show only slight differences from other TZP materials. Therefore, the performance of TZP ceramics produced from electrofused powders can be considered adequate for most applications.

4. CONCLUSIONS

The results reported, clearly indicate that the production of high performance TZP ceramics from electrofused, fine zirconia powders is a practical proposition. Furthermore, such powders have become a commercial reality.

Electrofused powders behave differently to those manufactured by alternative routes. The fundamental characteristic of high particle density, makes fused powders ideally suited to processes where solids loading is particularly important. Resultant high green densities produce a corresponding low fired shrinkage, minimising the risk of warpage or distortion during drying and firing.

The mechanical properties obtained for sintered samples can be considered acceptable for the majority of applications for TZP zirconias. Larger mean particle sizes result in marginally lower bend strengths and higher fracture toughness values.

REFERENCES

1. GARVIE, R. C., HANNINK, R. H. & PASCOE, R. T., *Nature,* **258**, 703, (1975).
2. REITH, P. H., REED, J. S. & NAUMANN, A. W., *Bull. Am. Ceram. Soc.,* **55**, 717, (1976).
3. GUPTA, T. K., BECHTOLD, J. H., KUZNICKI, R. C., CADOFF, L. H. & ROSSING, B. R., *J. Mat. Sci.,* **12**, 2421, (1977).
4. MAZDIYASNI, K. S., *Ceramics Int.,* **8**, 42, (1982).
5. VAN DE GRAAF, M. A. C. G. & BURGGRAAF, A. J., *Advances in Ceramics,* **12:** *Science and Technology of Zirconia II,* 744, (1983).
6. BLACKBURN, S., KERRIDGE, C. R. & SENHENN, P. G., *Advances in Ceramics,* **24,** 193, (1988).
7. BLACKBURN, S., SENHENN, P. G. & KERRIDGE, C. R., *Advances in Ceramics,* **24,** 211, (1988).
8. BLACKBURN, S., HITCHENER, M. P. & KERRIDGE, C. N., *1st International Conference on Ceramic Powder Processing Science,* Orlando, publ. The American Ceramic Society, **1(B)**, 864, (1988).
9. BLACKBURN, S., KERRIDGE, C. R. & SENHENN, P. G., *1st International Conference on Ceramic Powder Processing Science,* Orlando, publ. The American Ceramic Society, **1(B)**, 1011–18, (1988).
10. CHEN, Y. L. & BROOK, R. J., *Brit. Ceram. Trans. J.,* **88**, 7, (1989).
11. Japanese Industrial Standard JIS R1601, (1981); *Testing Method for Flexural Strength (Modulus of Rupture) of High Performance Ceramics,* Translated and published by the Japanese Standards Association.
12. QUINN, G. D. & MORRELL, R., *J. Am. Ceram. Soc.,* **74**, 2037, (1991).

Grain Boundary Modification of Ce-TZP by a Small Alumina Addition

J. WANG, H. P. LI,* R. STEVENS,* C. B. PONTON and P. M. MARQUIS

IRC in Materials for High Performance Applications & School of Metallurgy and Materials, The University of Birmingham, Birmingham, B15 2TT, U.K.
*School of Materials, The University of Leeds, Leeds, U.K.

ABSTRACT

The effects of a small Al_2O_3 addition (0·6 wt%) in influencing the microstructure and mechanical properties of a commercially available CeO_2 (15·7 wt%) stabilized tetragonal zirconia polycrystals (Ce-TZP) have been investigated. The Al_2O_3 additive reacts with the intergranular silica/silicate glassy phase to form crystalline mullite inclusions at the grain junctions in the sintered Ce-TZP. Consequently, the liquid phase-assisted densification process of the silica/silicate contaminated Ce-TZP was modified by the introduction of 0·6 wt% Al_2O_3. The average grain size of the Ce-TZP containing 0·6 wt% Al_2O_3 is smaller than that of the Al_2O_3-free Ce-TZP when both are sintered at temperatures in the range 1300 to 1600°C. The reduction in the amount of silica/silicate glassy phase at the grain boundaries and grain junctions in the Al_2O_3 doped Ce-TZP leads to an improvement in both the fracture strength and fracture toughness relative to the Al_2O_3-free Ce-TZP.

1. INTRODUCTION

CeO_2 stabilized tetragonal zirconia polycrystals (Ce-TZP) are the ceramic materials which exhibit the highest fracture toughness ever measured for brittle monolithic materials [1, 2, 3]. This is a consequence of the stress-induced transformation toughening associated with the autocatalytic tetragonal to monoclinic transformation which occurs in Ce-TZPs [4, 5]. As with other engineering ceramic materials, the mechanical properties of these highly toughened ceramics are affected strongly by their compositional and microstructural parameters, such as criteria content, sintered density, grain size, grain size distribution, and grain boundary characteristics [2, 6]. It is likely that the densification of many Ce-TZPs, but not all, is a liquid phase-assisted process at the sintering temperature, due to the presence of silica/silicate impurities in the starting materials [7, 8]. The occurrence of an intergranular silica-rich glassy phase has been observed using transmission electron microscopy (TEM) both in Ce-TZPs and in Y_2O_3 stabilized tetragonal zirconia polycrystals (Y-TZPs). As has been well established, the intergranular glassy phase has an important impact on the autocatalytic tetragonal to monoclinic transformation and therefore on the mechanical properties of these materials [4, 5].

Commercially available CeO_2 doped zirconia powders are currently manufactured via two very different processing routes, namely the electrofusion and refining of mixed zirconia and ceria powders, and coprecipitation using zirconium and cerium nitrates and/or chlorides as the starting materials [9, 10]. It has been shown that the CeO_2-doped zirconia powders manufactured via the electrofusion and refining route are highly sinterable at relatively low sintering temperatures. For example, Ce-TZPs with controlled microstructures and almost fully sintered density are readily obtainable by sintering the electrofused and refined powders at temperatures from 1300 to 1500°C for an appropriate period [8, 10]. By contrast, some coprecipitated CeO_2 doped zirconia powders are much less sinterable. A much higher sintering temperature is required to densify the coprecipitated powders. It has been demonstrated that the presence of silica/silicate impurities in the electrofused and refined zirconia powders plays an important role in their densification behaviour and affects the microstructure and mechanical properties of the sintered materials, although the exact mechanisms by which the impurities operate are not well established [8].

It is well known that alumina reacts with silica at temperatures above 1000°C, resulting in the formation of crystalline mullite [11]. Therefore, the addition of an appropriate amount of alumina to the silica/silicate contaminated zirconia powders will lead to a modification of both the microstructural evolution during sintering and the mechanical properties of the sintered materials. Thus, the aim of the present work is to investigate the effects of a small Al_2O_3 addition (0·6 wt%) on the sintering behaviour of a commercially available electrofused and refined CeO_2 doped zirconia powder which contains 0·2 wt% silica/silicate, as well as on the microstructure and mechanical properties of the sintered materials.

2. EXPERIMENTAL

15·7 wt% CeO_2 doped zirconia powder, which was manufactured via an electrofusion and refining route using zirconia and ceria powders as the starting materials, was supplied by Unitec Ceramics Limited, Stafford, England. Table 1 lists the composition analysis for this powder as provided by Unitec Ceramics Limited. The principal impurity in the as-received zirconia powder is silica, the presence of which is due to the fact that most zirconia powders derive originally from zircon ($ZrSiO_4$, zirconium silicate) and baddeleyite, which are the main mineral sources for zirconia. In order to compensate for the effects of the 0·2 wt% silica impurity in the as-received zirconia powder, the necessary amount of alumina additive can be worked out on the basis of the mullite composition (72 wt% alumina and 28 wt% silica). Therefore, 0·6 wt% is the approximate amount of Al_2O_3 required for this electrofused and refined zirconia powder. Ball milling, for 12 hours in propanol using zirconia milling media, was employed to mix the as-received zirconia powder and an appropriate amount of aluminium nitrate together. The milled suspension was then dried using a combination of IR heating and

Table 1. The composition analysis results (wt%) for as-received 15·7 wt% ceria-doped zirconia powder manufactured via an electrofusion and refining route

ZrO_2	HfO_2	CeO_2	SiO_2	CaO	Na_2O	K_2O	Al_2O_3	Fe_2O_3	MgO
82·02	1·70	15·70	0·20	0·04	< 0·1	< 0·01	0·13	0·09	< 0·05

heating on a hot plate. It was considered that calcination at low temperatures (400 to 600°C) prior to powder compaction, in order to decompose the aluminium nitrate to alumina, was not necessary because the overall aluminium nitrate content was so small. The dried powder was first compacted uniaxially in a steel die, 45 mm in diameter, at a pressure of 35 MPa, and then cold isostatically at 300 MPa. Green compacts of the as-received powder (Al_2O_3-free) were prepared using exactly the same compaction procedure. The green compacts were thermally treated at 600°C for 4 hours to decompose the aluminium nitrate to alumina, followed by sintering at 1500°C for 2 hours using heating and cooling rates of 3°C/min. Green compacts of both the doped and undoped Ce-TZPs were fired at temperatures from 1100 to 1600°C at intervals of 50°C, in order to study the sintering shrinkage and sintered density as a function of sintering temperature.

For the mechanical property evaluations, the sintered materials were cut into bars of 4 × 5 × 20+ mm using a diamond cutting wheel. The test bars were then polished to remove surface flaws. For the SENB toughness tests, a single notch of 400 μm in width was introduced using a diamond blade. Mechanical property tests for both the three point bend strength and SENB fracture toughness [12] (span: 20 mm and crosshead speed: 0·5 mm/min) were carried out using an Instron test machine (8501 type) equipped with a heating furnace capable of operating at temperatures up to 1500°C. For the high temperature tests, the test bars were heated up at a rate of 10°C/min to each test temperature and held at the test temperature for 1 hour prior to testing. The fracture toughness was also measured using the Vickers indentation technique with an indentation load of 490 N [13], for comparison with the measured SENB toughness. Average grain size measurements were made using the linear intercept method on polished and thermally etched surfaces, more than 150 grains were counted in each material. XRD, and both SEM and TEM equipped with EDX were employed to characterize the microstructures of the sintered materials.

3. RESULTS AND DISCUSSION

Figure 1 shows the sintering shrinkage as a function of sintering temperature for the as-received zirconia powder compacted uniaxially in a steel die at 35 MPa and then cold isostatically pressed at 300 MPa. Limited sintering shrinkage occurred at temperatures below 1100°C. However, there was a rapid increase in the sintering shrinkage in the temperature range 1100 to 1300°C. Little further densification was observed at temperatures above 1300°C. Figure 2 illustrates the sintered density as a function of sintering temperature for both the composition containing 0·6 wt% Al_2O_3 and the Al_2O_3-free composition. The sintered density of the Al_2O_3-free composition rises sharply over the temperature range 1200 to 1300°C. At 1300°C, the sintered density reaches a maximum. Sintering at a higher temperature results in a slight decrease in the sintered density. This is in agreement with the temperature dependence of sintering shrinkage, as shown in Figure 1. Similarly, the sintered density for the composition containing 0·6 wt% Al_2O_3 increases

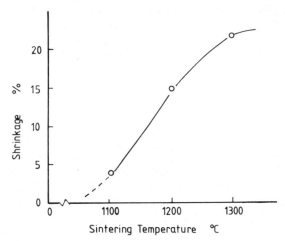

Figure 1. The sintering shrinkage as a function of sintering temperature for the as-received zirconia powder (Al₂O₃-free).

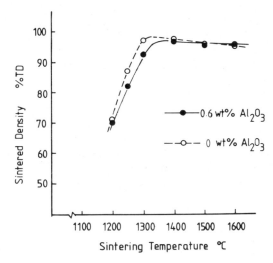

Figure 2. The sintered density as a function of sintering temperature for the compositions with and without a 0·6 wt% Al₂O₃ addition, respectively.

sharply with increasing sintering temperature over the same temperature range, although the increase is less than that of the Al_2O_3-free composition. Furthermore, the sintered density reaches a maximum on sintering at 1400°C, rather than at 1300°C. Sintering at 1250°C gives a sintered density of 82·2% and 86·7% theoretical density for the Al_2O_3-doped and Al_2O_3-free compositions, respectively. Greater than 98% theoretical density was achieved in the Al_2O_3-free composition at 1300°C. By comparison, the sintered density was 92% theoretical density for the composition containing 0·6 wt% Al_2O_3.

At temperatures between 1400 and 1500°C, the two materials exhibit similar sintered densities, although the Al₂O₃-free Ce-TZP is slightly more dense than the Al₂O₃-doped Ce-TZP. At 1600°C, the situation is reversed, *i.e.* the former is less dense than the latter. The slight drop in the sintered density of both materials at temperatures above 1400°C is due to the exaggerated grain growth, as discussed below.

As is shown in Figure 3, the average grain size of the Al₂O₃-free Ce-TZP is larger than that for the Al₂O₃ doped Ce-TZP over the entire sintering temperature range from 1300 to 1600°C. The difference in the average grain size between these two materials increases with increasing sintering temperature. This indicates that the addition of 0·6 wt% Al₂O₃ in the as-received zirconia powder inhibits grain growth on sintering at temperatures above 1300°C. Wang *et al.* [8] have established that the densification of the electrofused and refined zirconia powder is via a liquid phase-assisted process. The presence of 0·2 wt% silica/silicate impurity results in the formation of an intergranular liquid phase at the sintering temperatures, thereby promoting grain growth. The exaggerated grain growth results in a slight decrease in the sintered density, when the sintering temperature is above the temperature at which the maximum sintered density is achieved. However, the impurity effect of the intergranular silica/silicate phase will be partially or completely offset when 0·6 wt% Al₂O₃ is added into the as-received zirconia powder, by forming crystalline mullite inclusions at the grain junctions in the sintered material [11]. Furthermore, the occurrence of an intergranular crystalline mullite phase will inhibit grain growth by pinning the grain boundaries and/or grain

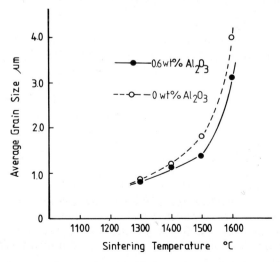

Figure 3. The average grain size as a function of sintering temperature in the temperature range from 1300 to 1600°C for the compositions with and without a 0·6 wt% Al₂O₃ addition, respectively.

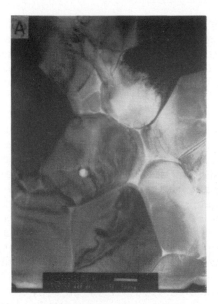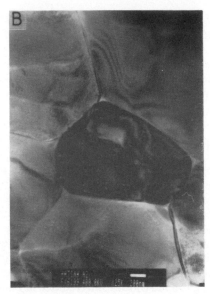

Figure 4. Bright field TEM micrographs showing the intergranular glassy phase which occurs in the Al₂O₃-free Ce-TZP sintered at 1500°C for 2 hours. EDX confirmed that the glassy phase is silicon-rich.

junctions [14]. The inhibited grain growth also explains the smaller decrease in the sintered density of the Al_2O_3-doped composition compared to the Al_2O_3-free composition on sintering at temperatures above 1500°C, as shown in Figure 2.

The formation of a crystalline mullite phase at the grain boundaries and grain junctions in the composition containing 0·6 wt% Al_2O_3 has been confirmed by microstructural studies using TEM equipped with EDX. Figure 4(a, b) are two bright field TEM micrographs showing the presence of an integranular silicon-rich glassy phase in the Al_2O_3-free Ce-TZP sintered at 1500°C for 2 hours. At the triple grain junctions, each zirconia grain exhibits a rounded morphology. Two or more triple grain junctions are linked together by a continuous glassy grain boundary layer, the thickness of which varies from a few nanometres to tens of nanometres. It was confirmed using TEM/EDX that these grain boundary phases are silicon-rich. Figure 5 shows two bright field TEM micrographs showing the microstructure of the sintered composition containing 0·6 wt% Al_2O_3. In contrast to the microstructure shown in Figure 4 for the Al_2O_3-free Ce-TZP, each zirconia grain is of angular morphology at the triple-grain junctions. The formation of a crystalline phase was observed at many, although not all, triple grain junctions. As shown in Figure 5, the composition of these intergranular crystalline phases is very close to that of mullite.

Figure 6 is a plot showing the three point bend strength as a function of temperature in air for both the Al_2O_3-free and the Al_2O_3-doped compositions.

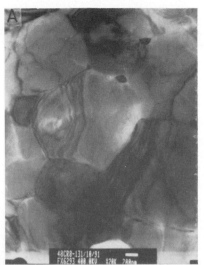

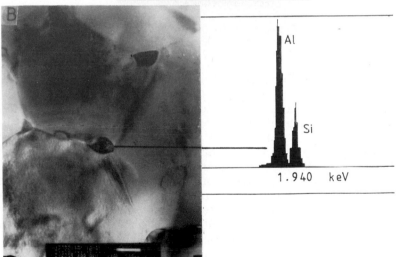

Figure 5. **Bright field TEM micrographs showing the crystalline secondary phases at grain junctions in the composition containing 0·6 wt% Al$_2$O$_3$. As indicated by EDX analysis result, the composition of these crystalline phases is very close to the composition of mullite (72 wt% Al$_2$O$_3$ + 28 wt% SiO$_2$).**

The fracture strength of the material containing 0·6 wt% Al$_2$O$_3$ is greater than that of the Al$_2$O$_3$-free composition at all test temperatures from room temperature up to 1200°C. With increasing test temperature, both materials exhibit a decrease in the three point bend strength from room temperature up to 600°C. A marked fall in the three point bend strength occurs in the temperature range of 600°C to 900°C. At temperatures above 900°C, only a

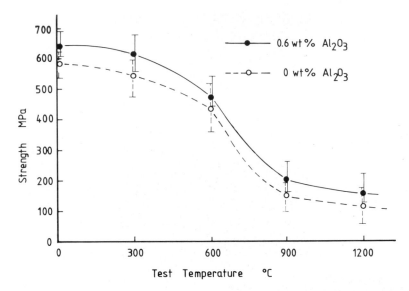

Figure 6. The three point bend fracture strength as a function of test temperature for the sintered materials with and without a 0·6 wt% Al₂O₃ addition, respectively.

quarter of the room temperature fracture strength is retained by the two materials.

Two principal parameters which affect the temperature dependence of fracture strength of transformation toughened ceramics are the starting temperature for the tetragonal to monoclinic transformation and the characteristics of any grain boundary phases if present [15, 16]. Thermodynamically, the free energy change associated with the tetragonal to monoclinic transformation in partially stabilized zirconia ceramics decreases with rising temperature. When the test temperature is equal to or above the tetragonal to monoclinic equilibrium temperature, which is dependent on parameters such as the type and amount of stabilizer and the grain size, the stress-induced transformation toughening will be thermodynamically impossible. The fall in fracture strength in the temperature range from 600 to 900°C, as shown in Figure 6, cannot be readily attributed to the softening of intergranular glassy phase, because the two materials exhibit a similar behaviour although they are considerably different with respect to the type and amount of intergranular phase present. It can therefore be concluded that the tetragonal to monoclinic transformation temperature in the 15·7 wt% CeO₂-doped TZP fabricated in the present work is in the temperature range of 600 to 900°C. The fall in the three point bend fracture strength over the temperature range of 600 to 900°C is due to the fact that it is impossible for stress-induced transformation toughening to occur at temperatures above the tetragonal to monoclinic transformation temperature.

The temperature dependence of the fracture toughness of the two materials has yet to be investigated. The fracture toughness values measured at room temperature using both the SENB and Vickers indentation techniques are given in Table 2. The Ce-TZP containing 0·6 wt% Al_2O_3 is slightly tougher than the Al_2O_3-free Ce-TZP. As mentioned earlier, the tetragonal to monoclinic transformation in Ce-TZPs is an autocatalytic process [4, 5]. The volume expansion and shear strain associated with the transformation in one grain may trigger the transformation in neighbouring grains. However, the existence of a glassy phase at the grain boundaries and grain junctions is likely to affect the autocatalytic nature of the tetragonal to monoclinic transformation. For example, an intergranular glassy phase will accommodate the residual stresses associated with the anisotropic thermal contraction of the tetragonal zirconia grains on cooling from the sintering temperature. Furthermore, the strain caused by the volume expansion and shear strain associated with the tetragonal to monoclinic transformation in one grain can also be partially or completely accommodated by an intergranular glassy phase, which has lower elastic moduli than the crystalline tetragonal zirconia grains.

Table 2. The SENB toughness and Vickers indentation toughness for the sintered Ce-TZPs with and without a 0·6 wt% Al_2O_3 addition, respectively

Materials	SENB, $MPa.m^{0.5}$	Indentation, $MPa.m^{0.5}$
Without Al_2O_3	9·87	10·6
With 0·6 wt% Al_2O_3	10·31	11·24

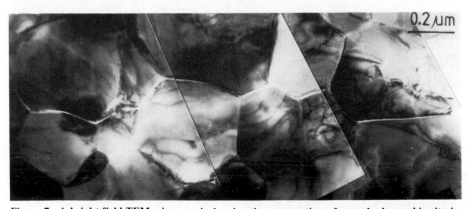

Figure 7. A bright field TEM micrograph showing the propagation of a crack observed in situ in an ion beam thinned Al_2O_3-free Ce-TZP foil. The crack went through a few grain boundaries and grain junctions without causing the transformation of any adjacent tetragonal grains.

Figure 7 is a bright field TEM micrograph showing the propagation of a crack observed in-situ on an ion beam thinned Al_2O_3-free Ce-TZP foil. The crack extends along a few grain boundaries and grain junctions without causing the transformation of any nearby tetragonal zirconia grains. This observation illustrates the importance of an intergranular glassy phase in affecting the nature of the tetragonal to monoclinic transformation in Ce-TZPs. The presence of the intergranular glassy phase, which is much more brittle and weaker than the tetragonal zirconia grains, reduces the possibility of nucleating the autocatalytic tetragonal to monoclinic transformation in nearby grains when the material is fractured. Figure 8 shows two SEM micrographs of fracture surfaces of test bars of the Al_2O_3-free Ce-TZP fractured at 600°C and 1200°C, respectively. At 600°C, the fracture is a mixture of intergranular and transgranular modes. In particular, the transgranularly fractured zirconia grains underwent the stress-induced tetragonal to monoclinic transformation, which is shown by the occurrence of twin steps, when the material was fractured at 600°C. By contrast, the fracture at 1200°C is almost 100% intergranular. When the material was fractured at 1200°C, the cracks went through the boundaries and junctions of the thermodynamically stable tetragonal grains.

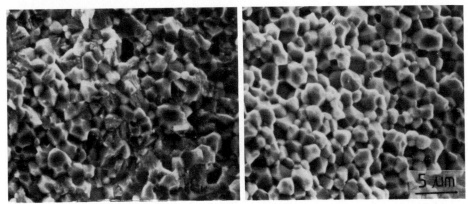

Figure 8. SEM micrographs showing the fracture surfaces of the sintered Al_2O_3-free Ce-TZP at (a) 600°C and (b) 1200°C, respectively.

4. CONCLUSIONS

A small Al_2O_3 addition (0·6 wt%) to a commercially available CeO_2 doped zirconia powder, which is manufactured via an electrofusion and refining route, results in a modification both in the sintering behaviour, and in the microstructure and mechanical properties of the sintered Ce-TZP. The Al_2O_3 additive reacts at the sintering temperature with the intergranular silica/silicate glassy phase, forming crystalline mullite inclusions at the grain junctions. As a consequence of both the partial elimination of the intergranular liquid phase at the sintering temperature and the grain boundary pinning by the resultant crystalline mullite inclusions, the average grain size in the Al_2O_3-doped Ce-TZP is smaller than that of the Al_2O_3-free Ce-TZP when both compositions are sintered at temperatures between 1300 and 1600°C. The reduction in the amount of intergranular glassy phase present in the Al_2O_3-doped Ce-TZP enhances the autocatalytic nature of the stress-induced tetragonal to monoclinic transformation. Therefore, the Al_2O_3-doped Ce-TZP exhibits an improvement in both the fracture strength and fracture toughness relative to the Al_2O_3-free Ce-TZP.

REFERENCES

1. EVANS, A. G. & CANNON, R. M., *Acta Metall.*, **34**, 761, (1986).
2. TSUKUMA, T. & SHIMADA, M., *J. Mater. Sci.*, **20**, 1178, (1985).
3. NETTLESHIP, I. & STEVENS, R., *Intl. J. High Tech. Ceram.*, **3**, 1, (1987).
4. CHEN, I. W. & REYES-MOREL, P. E., *J. Amer. Ceram. Soc.*, **69**, 181, (1986).
5. REYES-MOREL, P. E., CHERNG, J. & CHEN, I. W., *J. Amer. Ceram. Soc.*, **71**, 648, (1988).
6. WANG, J., RAINFORTH, W. M. & STEVENS, R., *Brit. Ceram. Trans. J.*, **88**, 1, (1988).
7. STEVENS, R., "An Introduction to Zirconia," 2nd Edition, MEL, Twickenham, (1986).
8. WANG, J., ZHENG, X. H. & STEVENS, R., Accepted for publication in the *J. Mat. Sci.*, in press, (1992).
9. TSUKUMA, T., *Amer. Ceram. Soc. Bull.*, **65**, 1386, (1986).
10. BLACKBURN, S., BLENHEIM, P. G. & KERRIDGE, C. R., in *Advances in Ceramics, **24A**, Science and Technology of Zirconia III*, Eds. S. Somiya, N. Yamamoto and H. Hanagida, The American Ceramics Society, Westerville, OH, (1988), pp. 211–15.
11. AKSAY, I. A., DABBS, D. M. & SARIKAYA, M., *J. Amer. Ceram. Soc.*, **74**, 2343, (1991).
12. BROWN, W. F. & SRAWLEY, J. E., *ASTM Special Technical Publication*, **410**, American Society for Testing Materials, (1967).
13. ANSTIS, G. R., CHANTIKUL, P., LAWN, B. R. & MARSHALL, D. B., *J. Amer. Ceram. Soc.*, **64**, 533, (1981).
14. KIBBEL, B. W. & HEUER, A. H., in *Advances in Ceramics, **12**, Science and Technology of Zirconia II*, Eds. N. Claussen, M., M. Ruhle and A. H. Heuer, The American Ceramics Society, Columbus, OH, (1984), pp. 415–24.
15. GARVIE, R. C., *J. Phys. Chem.*, **69**, 1238, (1965).
16. DAVIDGE, R. W., "Mechanical Behaviour of Ceramics," Cambridge University Press, London, (1979).

Compaction and Sintering of Toughened Zirconia Ceramics

J. L. HENSHALL
School of Engineering, University of Exeter, Exeter, Devon, EX4 4QF, UK
and S. T. THURAISINGHAM
Colloids-in-Industry, P.O. Box 263, Concord, MA 01742, USA

ABSTRACT

This investigation compares the indentation hardness and fracture results obtained on a commercial ceria stabilised tetragonal polycrystalline zirconia with those obtained by sintering green compacts formed either by uniaxial dry pressing or wet compaction. The purpose of the present programme is to develop the wet compaction process, which is inherently low lost and flexible, to achieve comparable properties to those obtained by dry pressing techniques, and without the large defects which exist in compacts formed by this route. The material manufactured by the wet compaction process has a similar, or slightly improved hardness, but is not as tough.

1. INTRODUCTION

Ceramics are now being considered by many design engineers as viable alternatives in many sophisticated and demanding technical devices, *e.g.* bearings, mechanical seals, machine tools and filters, where the prime requirements are resistance to wear and chemical attack. The wear resistance of ceramics is controlled by a combination of the hardness and toughness, and an estimate of the form of this dependence has been derived by Evans and Wilshaw [1]. It has been proposed [2] that the critical stress intensity derived from indentation fracture measurements is the most appropriate method of determining the toughness with respect to wear resistance, since the loading conditions are closer to the service conditions than in a conventional notched beam type test.

The interest in zirconia ceramics for load-bearing engineering situations stems from the work of Garvie *et al.* [3], who demonstrated that the toughness of zirconia ceramics could be dramatically improved by astute microstructural control utilising transformation toughening. Further refinements and developments of this approach [4, 5] have given rise to a range of zirconia ceramics with many potential applications. The main cubic/tetragonal phase stabilising agents used for these advanced zirconias are magnesia, yttria or ceria. The yttria stabilised materials have been shown to be prone to environmental degradation [6–9] and the magnesia stabilised zirconias require critical control of the annealing conditions and also powders of the MgO stabilised material undergo continuous leaching of the magnesium ions in an aqueous environment [10]. It has been shown that for ceria stabilised tetragonal zirconia, Ce-TZP, that provided the ceria content is greater than 10 mol%, low temperature degradation in air or water does not occur [11, 12]. Consequently, this study has concentrated upon Ce-TZP ceramics containing greater than 10 mol% ceria.

Several methods of powder compaction have been investigated for zirconias, *e.g.* uniaxial dry-pressing [13], cold high pressure isostatic pressing [14], hot isostatic pressing [15], slip casting [16] and tape casting [17]. The problems associated with dry pressing techniques are that only simple shapes can be produced by uniaxial pressing, isostatic pressing is expensive and large defects can exist in the sintered compact [18]. The aim of the present programme is to develop a technique using control of the surface chemistry and rheology of aqueous-based suspensions [19, 20] to prepare complex shaped compacts containing no large defects relatively inexpensively, and with other properties, in particular hardness and toughness, comparable to Ce-TZPs fabricated by alternative routes. This paper compares the hardness and toughness of a commercially available Ce-TZP with preliminary results obtained on similar ceramics produced using dry-pressing or water-based compaction, followed by sintering in air at 1400°C.

2. EXPERIMENTAL PROCEDURE

2.1 Materials and Processing

The starting powder used in this study was a 12 mol% ceria stabilised zirconia of median particle size 0·58 μm. An all glass still was used to prepare once-distilled water with specific conductance less than 2 μmhos-cm^{-1}. Twice distilled water was produced by a second distillation of the once-distilled water, containing additions of potassium permanganate and potassium hydroxide to remove traces of organic matter. A solution of Decon 90 of the recommended stength made with once-distilled water, followed by rinsing with twice-distilled water, was used to cleanse the preparation equipment. All solutions and suspensions were made using twice-distilled water. The powders were washed in once-distilled water prior to use until a constant conductance value of the rinse was obtained.

For dry pressing, the zirconia powder was mixed with either 1 or 2 wt% of binder, stearic acid, and lubricant, polyvinyl alcohol, for 30 mins. by hand prior to pressing into 10 mm diameter pellets in a hand operated hydraulic press at pressures of 54, 68, 79 or 85 MPa. The firing schedule was to heat to 500°C, hold for 2 hours, then fire at 1400°C in air for 1, 2 or 3 hours.

The full experimental details of the water-based compaction procedure cannot be presented since a patent application is being considered. The main parameters of interest however are the surface chemistry of the powder particles, the rheology of the suspension, which depends on the solids volume fraction and the particle surface state, and the compaction pressure. After this initial compaction the specimens were placed in a thin-walled rubber bag and isostatically pressed at 100 MPa. The specific surface areas of some of the compacts were measured using the nitrogen adsorption BET point B method, since this gives an indication not only of the total porosity levels, as would relative density, but also the pore sizes. The firing schedule in this case involved holding at 200°C for 2 hours prior to final sintering at 1400°C for 3 hours in air.

The commercially supplied Ce-TZP was in the form of tiles 60 mm × 60 mm × 5 mm. The CeO_2 content was determined as 10·5 mol% using a Link energy dispersive X-ray analysis with an Hitachi S520 scanning electron microscope. X-ray diffraction showed that the crystal structure was tetragonal with no detectable monoclinic or cubic zirconia phases. As reported previously [21], SEM examination of polished and etched specimens showed that the grain size was typically 1–2 μm, with larger grains (up to 5 μm) and occasional gross defects (greater than 0·1 mm).

2.2 Indentation Testing

The specimen surfaces were ground and polished using conventional metallographic procedures of water-lubricated SiC papers followed by polishing with 7 μm, 3 μm and 1/4 μm diamond paste impregnated cloths. Sufficient material was removed to ensure that any surface effects, *e.g.* residual stress, oxygen depletion, etc. would not influence the results. A Vickers diamond pyramidal indenter was used with loads in the range 294·3 N–882·9 N. The time taken for the full load to be applied was 10 s and the dwell time 20 s. Between four and six well-spaced indentations were made on the specimen's surface. The diagonals of the indentations, and the lengths of the cracks formed at each corner, were measured optically using a graduated eyepiece to an accuracy of 0·5 μm.

3. RESULTS AND DISCUSSION

The results of the sintering trials were generally successful. Since the specimens were relatively small, ca. 10 mm diameter × 4 mm, there were no problems with fracture on firing, although the resultant surface texture was generally quite rough. The surfaces of the specimens were not discoloured as a result of the heating.

As discussed above, the indentation test would appear to provide the most appropriate guide to wear resistance. Also, even with the present relatively limited size of specimen, sufficient tests can be performed to achieve good reproducibility. However, there is a problem concerning the analysis of the indentation cracking in these toughened ceramics. There are at least 20 formulae in the literature for deriving critical stress intensities from the indentation measurements [2], none of which is explicitly applicable to the present situation. It has been proposed [2] that since the indentation fracture situation involves crack arrest the symbol K_{Ia} should denote the derived parameter and also that the most suitable equations that can be used to analyse the data are [22]:

$$K_{Ia} = 0.4636 \frac{P}{a^{3/2}} \left(\frac{E}{Hv} \right)^{0.4} 10^F$$

where $F = -1.59 - 0.34x - 2.02x^2 + 11.23x^3 - 24.97x^4 + 16.32x^5$, in which $x = \log_{10}(c/a)$, where P is load, E = Young's modulus (taken as 210 GPa for all specimens), a is one half of the indentation diagonal, c is the apex-to-tip crack length, *i.e.* one half the tip-to-tip crack length, and Hv is the Vickers hardness at the test load.

or [23]:

$$K_{Ia} = 0.0312 \ (E^{0.4} \ P^{0.6}/a^{0.7}) \ (c/a)^Y \tag{2}$$

where $Y = (c/18a) - 1.51$.

or [24]:

$$K_{Ia} = 0.0154 \ (E/Hv)^{0.5} \ (P/c^{3/2}) \tag{3}$$

or [2]

$$K_{Ia} = 0.0195 \ (E/Hv)^{0.5} \ (P/c^{3/2}) \tag{4}$$

Table 1 lists the values of the indentation measurements on the commercial Ce-TZP for loads in the range 294·3 N to 882·9 N. The hardness is essentially constant in this load range at 8·0 GPa. The values of K_{Ia} however show a significant variation with load. The errors are typically $\pm$ 0·3 MPa m$^{1/2}$ [2], which is much less than the observed variation. The cracks are all Palmqvist type and the variations in K_{Ia} have been shown not to arise from residual surface stresses during manufacture or preparation [2]. The value of K_{Ic} for this material was determined as 7·7 MPa m$^{1/2}$ using 4-pt. bending of Single Edge Notched Beams. Equation (1) [21], gives values closest to this. However, this must be regarded as in part fortuitous since results for other ceramics show a distinct systematic variation with c/a when analysed using this formula [2]. Also, consideration of Equation (1) shows that for the values of c/a from Table 1 the coefficient F is virtually constant. Thus, since E and Hv are constant the critical stress intensity essentially depends on $P/a^{3/2}$, which cannot be constant if the hardness (which is P/a^2) is. The variations apparent using Equations (2)–(4) are consistent, which indicates that even though there is difficulty in ascribing absolute values to the toughness, at least there should be comparability, to enable the effects of processing variables to be discriminated.

Table 1. Variation of Vickers indentation half-diagonal length, a (μm), half tip-to-tip crack length, c (μm), hardness, Hv (GPa), and critical stress intensity factor K_{Ia} (MPa m$^{1/2}$), as derived from Equations 1–4, for commercially obtained Ce-TZP

				K_{Ia} (MPa m$^{1/2}$)			
Load (N)	a (μm)	c (μm)	Hv (GPa)	Eqn. 1	Eqn. 2	Eqn. 3	Eqn. 4
294·3	130	142	8·1	8·4	14·8	13·7	17·4
490·5	169	191	8·0	9·4	15·9	14·7	18·6
588·6	185	203	8·0	9·9	17·3	16·1	20·4
686·7	201	229	7·9	10·1	17·0	15·7	19·9
784·8	214	229	7·9	10·8	19·3	17·9	22·7
882·9	228	259	7·9	10·8	18·2	16·8	21·3

Table 2. Results for uniaxially dry-pressed and sintered (at 1400°C) compacts of Ce-TZP. PP is the pressing pressure in MPa, WBL is the weight% binder, which is also the same as the weight% of lubricant, ST is the sintering time (h), and the other symbols are defined in Table 1. n.a. indicates the result is not available

| | | | | | | K_{Ia} | | | |
PP	WBL	ST	a	c	Hv	Eqn. 1	Eqn. 2	Eqn. 3	Eqn. 4
54	1	2	148	161	6·3	7·7	13·5	12·8	16·3
68	1	2	146	166	6·4	7·6	12·8	12·1	15·4
79	1	2	142	168	6·8	7·6	12·3	11·6	14·7
85	1	2	141	174	6·9	7·5	11·7	10·9	13·8
54	1	1	145	181	6·5	7·3	11·3	10·6	13·4
54	2	1	146	157	6·4	7·8	13·8	13·2	16·7
54	1	3	128	n.a.	8·3				
54	2	3	140	n.a.	7·0				

The crack tip lies well within the plastic/transformed zone around the indentation, and therefore the mechanics of the situation will be quite different from that of crack propagation from a macroscopic notch when the transformation is limited to a few micrometres either side of the crack tip. Thus, it is probable that the higher values derived from Equations (2)–(4) are a more accurate reflection of the materials' resistance to cracking under these type of conditions in service than the SENB test.

Table 2 lists the values of indentation hardness and critical stress intensities, normal load 294·3 N, obtained on the compacts prepared by uniaxial dry pressing prior to sintering. The first four results show the effect of altering the pressing pressure for a given concentration of binder and lubricant and sintering time at 1400°C. As can be seen the hardness increased and the toughness decreased as the pressing pressure is raised. Higher values of pressure were also tried, but in these cases the sintered compacts either cracked or spalled. Thus, it was decided to investigate the effects of varying the wt% of binder/lubricant and sintering time using the lowest pressing pressure. Increasing the sintering time to 3 hours for 1 wt% of binder/lubricant had the most beneficial effect on the hardness. Increasing the sintering time also increased the toughness. The maximum values of hardness and toughness, 8·3 GPa and 13·8 MPa m$^{1/2}$, are comparable to those obtained on the commercial material, but they did not occur in the same compact.

Table 3 presents the results for the compacts formed via the wet pressing procedure. Four different pressing pressures were used in the ratio 1:2·71:2·86: 5·23, three different surface chemical conditions A, B and C, and two solid: liquid ratios. The effective viscosities of the mixes were dependent on both the surface chemistry and solid:liquid ratio. All the specimens were sintered for 3 hours at 1400°C after holding for 2 hours at 200°C. For surface chemistry A increasing the pressure ratio from 1 to 2·71 or 2·86 increased the hardness significantly, and also the toughness, of the compact. Changing the surface chemistry from A to B produced a marked reduction in the specific surface

Table 3. Effect on pre-compaction conditions of the properties of the sintered compacts. PR is the ratio of the applied pressures, SR is the ratio of the solids in a given volume of liquid, SC represents the different surface chemical conditions (which are controlled by the liquid compositions), SSA is the specific surface area of the green compacts (m^2/g {determined using the nitrogen adsorption BET point B method}), and the other green symbols are defined in Table 1

							K_{Ia}			
PR	SR	SC	SSA	a	c	Hv	Eqn. 1	Eqn. 2	Eqn. 3	Eqn. 4
1	1	A	n.a.	136	192	7·4	7·2	9·9	9·1	11·5
2·71	1	A	240	129	170	8·2	7·7	11·3	10·3	13·1
2·86	1	A	155	130	181	8·0	7·5	10·5	9·5	12·1
2·86	1	B	23	128	167	8·4	7·8	11·6	10·5	13·3
5·23	1	B	n.a.	127	176	8·5	7·6	10·6	9·6	12·2
2·86	1	C	n.a.	129	168	8·2	7·8	11·5	10·5	13·3
2·86	3	C	172	130	170	8·0	7·7	11·4	10·4	13·2

area of the green compact, suggesting a much finer distribution of porosity. The hardness, and possibly also the toughness, of the sintered compact was increased with respect to similar conditions for surface chemistry A. Further increasing the pressing pressure resulted in a negligible change in hardness and a slight reduction in K_{Ia}. The other surface chemical condition C, resulted in a slightly reduced hardness and comparable toughness to the values obtained for B at the same pressure ratio, *i.e.* 2·86. Increasing the solids content by a factor of three produced little effect on the compact properties. The hardness values obtained are slightly greater than those for the commercial Ce-TZP, but the K_{Ia} results are substantially less.

The values of hardness obtained in this study are less than reported by Tsukuma and Shimada [14], but greater than those of Duh *et al.* [13] for 12 mol% ceria stabilised zirconias sintered at 1400°C. Similar coprecipitation techniques were used in both cases, followed by dry pressing at ca. 40 MPa, but the former authors [14] subsequently isostatically pressed at 300 MPa, prior to sintering. Also the latter authors [13] reported that the hardness decreased with increasing sintering time, whereas the present results would generally suggest the opposite trend, at least up to 3 hours. Indentation critical stress intensity values presented in reference [13] were calculated using the equation of Niihara *et al.* [24], which in this regime gives toughness values ca. 50% greater than Equation (2) [2]. The values obtained for similar compositions and sintering times were significantly less than obtained in the present work. These authors also found no correlation between indentation toughness and the macroscopic K_{Ic} determined using chevron notched beams.

4. CONCLUSIONS

The main conclusions of this work are that reasonably good and consistent properties can be obtained for a variety of processing conditions using fine-sized 12 mol% ceria stabilised zirconia. The best conditions for dry pressing from the present results would appear to be a relatively low pressing pressure (ca. 50 MPa) followed by sintering for 3 hours at 1400°C. The conditions to use to produce the best compacts by wet pressing, *i.e.* surface chemistry B, pressure ratio ca. 2·86, and probably a wide latitude for the solids content, were identified. The hardness of the fired compact was good, but the K_{Ia} values are relatively low. Further optimisation of the firing schedule is obviously required but the inherent flexibility and relatively low cost of the process suggest that further developments of the process are worthwhile.

ACKNOWLEDGMENTS

The authors gratefully acknowledge the technical assistance of Breman Thuraisingham and William Furniss.

REFERENCES

1. EVANS, A. G. & WILSHAW, T. R., *Acta Metall.,* **24**, 939, (1976).
2. GUILLOU, M.-O., HENSHALL, J. L., HOOPER, R. M. & CARTER, G. M., "Indentation Fracture Testing and Analysis and its Application to Zirconia, Silicon Carbide and Silicon Nitride Ceramics," presented at the Fourth International Conference on the Science of Hard Materials, Madeira, Nov., (1991), accepted for publication in the *Journal of Hard Materials.*
3. GARVIE, R. C., HANNINK, R. H. & PASCOE, R. T., *Nature,* **258**, 703, (1975).
4. CLAUSSEN, N., *Mater. Sci. Eng.,* **71**, 23, (1985).
5. BECHER, P. F., *J. Amer. Ceram. Soc.,* **74**, 255, (1991).
6. KOBAYASHI, K., KUWAJIMA, H. & MASAKI, T., *Solid State Ionics,* **3–4**, 489, (1981).
7. SATO, T. & SHIMADA, M., *J. Amer. Ceram. Soc.,* **67**, C212, (1984).
8. SATO, T. & SHIMADA, M., *J. Amer. Ceram. Soc.,* **68**, 356, (1985).
9. SATO, T., OHTAKI, S. & SHIMADA, M., *J. Mater. Sci.,* **20**,1466, (1985).
10. CARIER, G. M., University of Exeter, private communication.
11. SATO, T. & SHIMADA, M., *Amer. Ceram. Soc. Bull.,* **64**, 1382, (1985).
12. MATSUMOTO, R. L. K., *J. Amer. Ceram. Soc.,* **71**, C128, (1988).
13. DUH, J. G., DAI, H. T. & CHIOU, B. S., *J. Amer. Ceram. Soc.,* **71**, 813, (1988).
14. TSUKUMA, K. & SHIMADA, M., *J. Mater. Sci.,* **20**, 1178, (1985).
15. KIM, J. Y., UCHIDA, N., SAITO, K. & UEMATSU, K., *J. Amer. Ceram. Soc.,* **73**, 1069–73, (1990).
16. MORENO, R., REQUENA, J. & MOYA, J. S., *J. Amer. Ceram. Soc.,* **71**, 1036, (1988).
17. RICHARDS, V. L., *J. Amer. Ceram. Soc.,* **72**, 325, (1989).
18. ULRICH, D. L., *Chem. Eng. Proc.,* **68**, 28, (1990).
19. WAKEMAN, R. J., THURAISINGHAM, S. T. & TARLETON, E. S., *Filtration and Separation,* **26**, 277, (1989).
20. LEONG, Y. K. & BOGER, D. V., *J. Rheology,* **35**, 149, (1991).
21. CARTER, G. M., HOOPER, R. M., HENSHALL, J. L. & GUILLOU, M.-O., *Wear,* **148**, 147, (1991).
22. EVANS, A. G., *ASTM STP No. 678, American Society for Testing and Materials,* Philadelphia, PA, USA, 112, (1979).
23. LIANG, K. M., ORANGE, G. & FANTOZZI, G., *J. Mater. Sci.,* **25**, 207, (1990).
24. ANSTIS, G. R., CHANTIKUL, P., LAWN, B. R. & MARSHALL, D. B., *J. Amer. Ceram. Soc.,* **64**, 533, (1981).

Influence of the Addition of Solid Solution MgO on the Microstructure of Ce-TZP

XIUHAU ZHENG

Department of Mechanical Engineering, Beijing Institute of Technology, Beijing 100081, P.R. China

NIEVES GARCÍA CORONADO and R. STEVENS

School of Materials, The University of Leeds, Leeds, LS2 9JT, U.K.

ABSTRACT

The influence of MgO additions to Ce-TZP powder on the sintered microstructure of the ceramic has been studied. In the absence of MgO the microstructure consists of the tetragonal phase with only a small fraction of the monoclinic form present. The predominant characteristic features of this microstructure were the presence of isolated dislocations and stacking faults and on rare occasions arrays of dislocations in the form of small-angle grain boundaries. The t $\rightarrow$ m martensitic phase transformation tended to occur with the formation of isolated needles, very often inducing autocatalytic reaction in adjacent grains, which could be observed using TEM. The morphology of the transformation product is believed to be due to a minimum strain energy and shape change requirement.

When MgO was added to this material the microstructural features changed dramatically, with an increased number of grains having a relatively heavy density of dislocations, and frequent arrays of dislocations forming substructures within the grains. The presence of monoclinic twins (even near microcracks) was seldom observed. Two different types of as yet unidentified features have also been seen in considerable numbers. It is suggested that the presence of such features is reflected in a change in the transformation behaviour which will indirectly influence the mechanical properties. Consequently, the control of the mechanical properties by thermal management of the microstructure is anticipated as being possible.

1. INTRODUCTION

In the last few years, considerable attention has been devoted to zirconia ceramics and more recently much of this has been focussed on the ceria-zirconia system. The reasons for this attention is the unusual combination of mechanical properties and in particular the enhanced toughness that can be developed in this ceramic.

For structural applications, there are two main groups of zirconia ceramics that have been developed, tetragonal zirconia polycrystals (TZP), and partially stabilized zirconia (PSZ). The former group utilises two main compositions, Y-TZP which was developed first and subsequently the Ce-TZP based compositions. Even though both have a similar phase composition, they demonstrate considerably different mechanical properties, Y-TZP exhibiting a higher strength than Ce-TZP, but a lower toughness [1]. Moreover, Ce-TZP appears to undergo a pseudo-plastic deformation prior to failure [2], a feature which is rarely found in the failure of the Y-TZP ceramic.

The microstructure of Y-TZP including the morphology and crystallography of its martensitic transformation has been widely studied. The microstructure of Ce-TZP and its martensitic $t \rightarrow m$ transformation has not yet been fully characterised, but is the subject of intense investigation.

The PSZ group of ceramics has several compositions which are commercially exploited, the most popular being MgO-PSZ, mainly due to the ability of this system to develop a range of properties through thermal treatments [1].

MgO-PSZ can be produced to develop a relatively high toughness and strength, although the maximum values are lower than the individual toughness measured for Ce-TZP and for the strength of Y-TZP respectively.

The work presented in this paper is part of a current project which is focussed on the possibility of combining in one ceramic the advantages of MgO-PSZ with those of Ce-TZP, *i.e.,* a high toughness, acceptable strength and the ability of the system to generate further property enhancement through thermal treatments. With this in mind we have studied the influence of MgO additions in the solid solution range on the microstructure of Ce-TZP, and the tetragonal to monoclinic martensitic transformation of this composition.

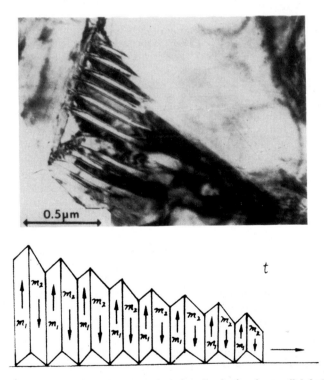

Figure 1. Micrograph showing the monoclinic twins developing in parallel fashion (a) and a diagram of the structure (b).

2. EXPERIMENTAL PROCEDURE

The materials used in this study were produced by a common route but with different starting compositions. The standard material contained 12 mol% of ceria (Powder A) and the other was a mixture of MgO with ceria (Powder B). After mixing, milling and pressing the samples into 25 mm discs, the green powder discs were pressureless sintered at temperatures ranging from 1400 to 1550°C to produce ceramics within a range of controlled grain size. The grain size of the surface was measured by SEM, and the phase content using XRD.

The phase content on the ground surfaces of the sintered discs was found to be different. Powder A showed a higher percentage of monoclinic after sintering at 1400°C; this decreased when it was sintered at 1500°C. Powder B exhibited very little monoclinic phase even when sintered at 1400°C.

The lattice parameters for Powder A were taken from the literature [3] whereas the lattice parameters of Powder B were determined from X-ray measurements and found to be somewhat different as outlined below. When the work with Powder A was done, data for the tetragonal phase was obtained from the card JCPDS 17-923. Subsequently a new card (JCPDS 38-1437) was published. The new card shows similar values of the lattice parameters to those obtained for Powder B (a = 3·6377, c = 5·2394 Å), the minor differences between the reference card and the measured results is considered to be due to compositional variation, since the powder under examination contains CeO_2 in solid solution as the stabilizing agent for the tetragonal phase.

Lattice parameters	Powder A nm	Powder B nm
Monoclinic		
a	0·512	0·536
b	0·518	0·522
c	0·530	0·520
β	81·1	98·88
Tetragonal		
a	0·512 (JCPDS 17-923)	0·362
c	0·522	0·521
a	0·36377 (JCPDS 38-1437)	
c	0·52394	

3. MARTENSITIC TRANSFORMATION CHARACTERISATION

Characterisation, using TEM, was carried out on sample A, and showed specific features. The monoclinic phase was invariably heavily twinned, with four typical variants frequently occuring in this material:

(a) Parallel twin-related variants which usually appeared to originate at or from the grain boundary, Figure 1.

(b) A "zig-zag" shape with a conjunction plane parallel to a twin plane, Figure 2.

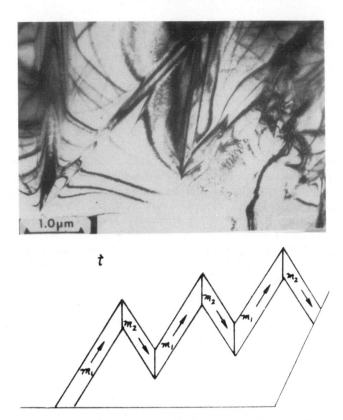

Figure 2. Micrograph showing the monoclinic twins in (a) "zig-zag" mode; (b) a corresponding diagram of the structure.

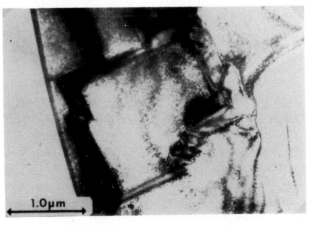

Figure 3. Micrograph showing a network of twins.

(c) A network of twinned variants within the grain, Figure 3.

(d) Individually nucleated martensitic plates in a specific twin relationship.

Linear and planar defects were detected within some of the monoclinic laths, somewhat akin to microtwinning; antiphase domain boundaries and dislocations were also present in some of the crystals.

3.1 Nucleation and Formation of Monoclinic Phase

The monoclinic phase in the form of laths or plates tended to nucleate preferentially at the grain boundaries, most frequently near triple or tetradic points, where it is thought the highest stress intensity is generated from thermal expansion mismatch, more specifically from thermal expansion anisotropy of adjacent tetragonal grains. It was also found that heterogeneous nucleation of the transformed phase could sometimes occur at small-angle grain boundaries, but this was less prevalent.

The $t \rightarrow m$ transformation behaviour exhibited by this material appears to be autocatalytic. The stresses and strains generated by the transformation event in one grain appear to induce further nucleation events either on the t/m interface within the grain or in other adjacent grains.

Such observations strongly suggest that the nucleation barrier in this material is critical. The martensitic transformation occurs only when stress state of the material is such that it undergoes the $t \rightarrow m$ transformation; the development of the "zig-zag" morphology (Figure 2) is the result of an insufficient level of stress to nucleate another further variant in the parallel fashion (Figure 1). The "zig-zag" morphology which results is the consequence of the minimum strain energy condition.

Complex networks of twins were produced to absorb the necessary change of shape required by the deformation which takes place during the transformation event, with a net shape strain which approaches zero, by self-accommodation of the twins. Small twins are formed within the network of the transformed product phase to accommodate the accumulated strain energy (Figure 3).

It is generally accepted that the transformability of zirconia based ceramic materials (TZPs) depends on the grain size of the tetragonal phase and the amount of stabilizer added [1], and that larger grains transform more easily than smaller ones. It is believed that the critical size condition arises from the domination of the transformation kinetics by difficulty of nucleating the monoclinic phase and that the larger grains are more likely to contain more effective nucleation sites [4].

In the present observations, the results turned out unexpectedly, to be different from this prediction, for transformation in the specimens with the larger grain sizes appeared to be considerably more difficult than for the smaller grain sized material. Such behaviour could be interpreted in terms of nucleation kinetics. It is generally believed that the nucleation of the $t \rightarrow m$ martensitic transformation is invariably stress-induced. Previous work has shown that the spinodal shear strain must be reached over a critical volume within a grain [5, 6] for the transformation to initiate.

The samples with the smaller grain size have higher density of grain boundaries and of triple points per unit volume, and thus contain more nuclei sites per unit volume, which statistically makes more likely the chance of a transformation event taking place.

A strong tendency was noted for smaller grain size material to transform to parallel variants, whilst in the specimens with larger grain size the "zig-zag" shape of the monoclinic transformation product and single plates was more frequently observed, which suggests that re-nucleation is easier in smaller grains compared with larger grains. This could well be due to a higher level of internal strain being retained in the finer grain material.

4. MICROSTRUCTURAL CHARACTERISATION AND INFLUENCE OF MgO

Samples A and B did not show any monoclinic phase to be present at the beginning of the TEM session, but sample A was seen to transform during examination in the microscope, whereas material B did not transform as easily. This was always the case, even under continuous exposure to the electron beam, or in regions near to microcracks, where high stresses would be expected and transformation likely. When material B does transform it seems to do so in the same fashion as sample A.

The detailed microstructure of the two samples differed somewhat, sample B having several features which did not appear on the magnesia free sample:

Fine "precipitates" were found within the grains. They appear to occur into two different size ranges, the first one from 4 nm to 10–15 nm, and the second from 20 to 40 nm (Figures 4 and 5).

The formation of substructures within both the tetragonal and cubic grains by networks of dislocations is common. The matrix can be seen to be highly strained (Figure 5) as evidenced by the heavy contrast present in the micrographs.

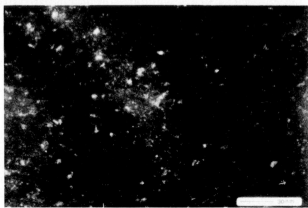

Figure 4. Micrograph showing "fine precipitates" in the matrix.

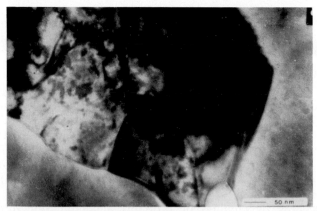

Figure 5. Micrograph showing contrast typical of the "coarse precipitates."

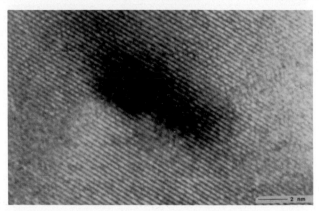

Figure 6. Lattice image showing how the planes bend when they go through the contrast zone produced by the "precipitate."

4.1 "Precipitate" Characterisation

Direct observations of the fine "precipitates" was very difficult due to the small size of the features and the presence of strains and dislocations in the microstructure. Consequently it was not possible to focus such features every time they appeared. When some of these features were oriented in the desired diffraction condition we could not observe any precipitate, but some contrast was seen which appeared to be due to a small mismatch of the planes within the grain (Figure 6). This is consistent with the difficulty in identifying them from the matrix using the SAD. Such features have been seen in monoclinic and tetragonal grains.

The coarser "precipitates" show similar behaviour to the smaller ones, *i.e.* it has not been possible to identify them as being categorically different from the matrix. The observed contrast is of a different shape and size to that of the fine precipitates and is also observed in the monoclinic as well as the tetragonal grains. It would appear that there is some difference between the two observations, although the nature of the effects producing the contrast has yet to be determined.

4.2 Substructures

The fact that the grains contained extensive substructures (Figure 5) suggests that any dislocations present in the matrix have been mobile at the sintering temperature. The quantity of such dislocation networks present is unusual for a zirconia ceramic. The arrays can be present as a two dimensional network of dislocations in the form of a twist/tilt sub-boundary (Figure 7) or as a simpler regular array of dislocations (Figure 8). The grains often exhibit regions of high contrast due to high elastic strains. It is not clear if the strains are produced by twinning or whether there is another source of strain (*e.g.* thermal mismatch on cooling) or probably a combination of both these effects.

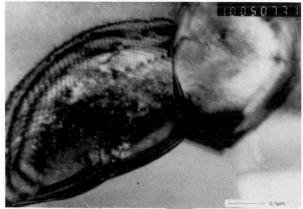

Figure 7. Micrograph showing dislocation arrays in two dimensions.

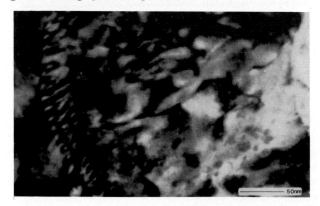

Figure 8. Micrograph showing dislocation arrays in a subgrain border together with strain fields of other dislocations in the adjacent grain.

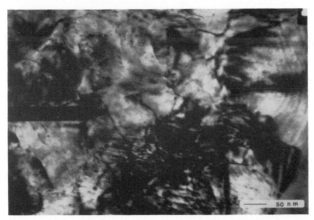

Figure 9. Micrograph showing numerous "pinned" dislocations.

In metallic systems the dislocations are rearranged to produce subgrain structures due to the lower energy of this distribution. Such recovery processes usually occur on annealing after heavy plastic deformation. The original defects present in the zirconia ceramics could well arise during the powder production process (exceptionally heavy milling). The low sintering temperature, in conjunction with the high impurity content, is also aided by the development of a glass phase formed by the impurity elements, which tends to form at the sintering temperature. This would tend to pin any dislocation produced by a form of "ageing" mechanism where impurity ions migrate to and settle on dislocation cores (Figure 9) thereby minimising the charge imbalance and lowering the overall elastic strain of the defect.

4.3 Deformation

The most interesting feature of this material is its ability to absorb "deformation." The primary relief process around microcracks appears to be strain fields within grains, some twinning and the presence of isolated dislocations and arrays. The very fine microcracks often observed at the end of twin laths in the Y-TZP were not present in this material which indicates that high stresses were either avoided or accommodated in some manner.

5. SUMMARY

(1) The nature of the martensitic transformation in Ce-TZP has been shown to differ if MgO is added to the ceramic to form a solid solution.

(2) Additions of MgO affect the microstructure creating and stabilising features such as a highly strained microstructures and arrays of dislocations.

REFERENCES

1. STEVENS, R., *An Introduction to Zirconia,* Magnesium Elektron Ltd., Publication No. **113,** Manchester, (1983).
2. TSUKUMA, K. & SHIMADA, R., *J. Mat. Sci.,* **20,** 1178, (1985).
3. JCPDS Card No. 17-923.
4. WAYMAN, C. M., in *Solid-Solid Phase Transformations,* Eds. H. I. Aaronson, David E. Langhlin, R. F. Sekerka and C. Marvin Wayman, publ. Metallurgical Society of AIME, New York, 1119, (1982).
5. CLAPP, P. C., *Physica Status Solidi,* **57,** 561, (1973).
6. SUZUKI, T. & WUTTING, M., *Acta Met.,* **23,** 1069, (1975).

The Formation of Al₂O₃/Al Composites by Oxidation

PING XIAO and B. DERBY

Department of Materials, University of Oxford, Parks Road, OX1 3PH, Oxford

ABSTRACT

Al_2O_3/Al composites can be produced by the direct oxidation of Al (DIMOX process) by oxygen in air at a certain range of temperatures. In this work we concentrate on the investigation of this process for pure Al with additions of MgO powder on the top surface exposed to air. Thermal gravimetric analysis is used to characterise the reaction kinetics. Optical microscopy and scanning electron microscopy coupled with EDX and WDS microanalysis are used to characterise the resulting microstructures. The experimental results are consistent with the mechanism of a proposed DIMOX process for MgO/Al reaction.

1. INTRODUCTION

Al_2O_3/Al composites can be produced from the oxidation of some molten Al alloys or pure Al metal with the addition of certain surface dopants over a limited temperature range (DIMOX process, Lanxide company, US). The oxide reaction product grows outwards from the original metal surface. The reaction is sustained by the wicking of liquid metal through continuous channels in the product to form an oxide/metal composite [1]. The microstructure of the composite, so produced, is an intimate mixture of continuous interconnected networks of ceramic and metal.

Initial studies considered the formation of Al_2O_3/Al composites by the oxidation of molten Al-Mg-Si alloys. The presence of both Mg and Si in the alloy was considered necessary for the reaction, [1, 2]. Since then simple Al-Mg alloys have been used for DIMOX experiments but in this case reaction is initiated by scratching the top surface of the alloy exposed to air prior to heating [3]. Nagelberg [4, 5] reported a study of the kinetics of oxidation of an Al-Mg-Si alloy and more complex Al-Si-Zn-Cu-Fe-Mg alloy which produced conflicting results. Using a combination of DTA and TGA techniques he estimated an activation energy for oxidation to be 400 kJ mol⁻¹ with dependence of the reaction rate on oxygen partial pressure of $p(O_2)^{1/4}$ for the Al-Mg-Si alloy, and an activation energy of 89 kJ⁻¹ with no effect of oxygen partial pressure for the Al-Si-Zn-Cu-Fe-Mg alloy. He has also proposed a model for the DIMOX process of Al-Mg alloy based on the kinetics of Al-Mg-Si [9]. It is supposed that there is a continuous MgO film on the top surface of the molten alloys, with liquid Al beneath the MgO film and Al_2O_3 below the liquid Al. The transportation of oxygen to the Al_2O_3 below the liquid Al is by diffusion of oxygen through the MgO film and liquid Al.

Al-Mg-Zn plus MgO powder on Al and NaOH on Al can all be used to produce Al_2O_3/Al composites [6, 7, 8]. The presence of Si is not required. The function of Mg during the process was considered in a model of the process. The present work continues research on the DIMOX process of MgO/Al to complete the model. Thermal gravimetric analysis was used to characterise reaction kinetics. Optical microscopy and scanning electron microscopy coupled with EDX and WDX microanalysis are used to characterise the resulting microstructures. The experimental results are explained by the mechanism we propose.

3. EXPERIMENTAL

Oxidation experiments were carried out using 99·99 wt% purity Al. Short cylindrical specimens of diameter 15 mm and height 16 mm were machined and inserted into crucibles of 99·9 wt% purity Al_2O_3 with an internal diameter slightly larger than the aluminium. MgO was deposited in powder form on the upper surface of the cylindrical aluminium (MgO/Al). The specimen was hung from a nichrome wire in a tube furnace at 1200°C and 1300°C in air. The specimen's weight was continually monitored using a laboratory microbalance with 0·1 mg accuracy and datalogged on a small microcomputer. A continuous record of oxidation rate was made during the reaction. These partially oxidised specimens were then sectioned for optical microscopy and microanalysis.

4. RESULTS

During the oxidation of pure Al with MgO powder on the top surface exposed to air, the MgO powder gradually decreases in amount and then disappears. After the disappearance of MgO powder, the oxidation continues. Microstructures from the top, the middle and the bottom of the composite are shown in Figure 1. These correspond to regions of the product formed during the final, intermediate and initial stages of oxidation respectively (the oxidising surface is that exposed to air at the top). The channel thickness of Al (white area) at the top of the specimen (1a) is smaller than that at the bottom (1c). In the middle there is a transition between the two areas with different thicknesses of channel. It is assumed that the MgO powder is used up at this interface. When the top piece of this product is cut off and the bottom one is put back in to the furnace at 1200°C again, oxidation continues and Al_2O_3/Al composites grow from the previous product. Figure 2 shows the interface area of an Al_2O_3/Al composites from this two stage oxidation. The difference of channel thickness in the two stage products (the right hand part is formed by the second stage reaction) is very clear. Figure 3 shows the microstructures of the products from MgO/Al mixtures with different sizes of MgO powder.

The kinetics of oxidation of MgO/Al was determined for different amounts of MgO on the top surface of Al [8]. From MgO = 4 mg/cm² to 31 mg/cm² on the top surface, the gradient of weight gain vs. time increases with the amount of MgO on the top surface. When different sizes of MgO powder but the same amount (20 mg/cm²) are applied to the top surface of the Al, the weight gain vs. time is as shown in Figure 4. The gradient is almost the same early in the process (at the first stage) but a little different later (at the second stage). The influence of temperature on weight gain vs. time is shown in Figure 5. Early in the oxidation, the weight gain at 1300°C is faster than that at 1200°C. At a later period the weight gain at 1200°C carries on until all the Al below the product is exhausted but the weight gain at 1300°C is almost zero

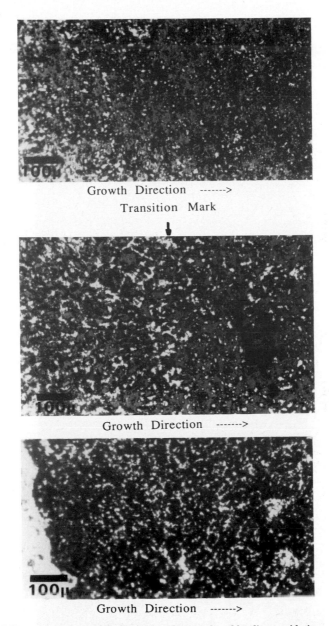

Growth Direction ------->

Transition Mark

Growth Direction ------->

Growth Direction ------->

Figure 1. Microstructures of Al₂O₃/Al composites produced by direct oxidation of pure Al doped with 20 mg cm⁻² MgO at 1200°C. (a) material near the air/ceramic interface (later forming the oxidation product), (b) middle region of the ceramic. The arrow indicates the transition from the material with large metal channels to that with small channels, (c) material close to the metal/ceramic interface (early oxidation product).

Transition Mark

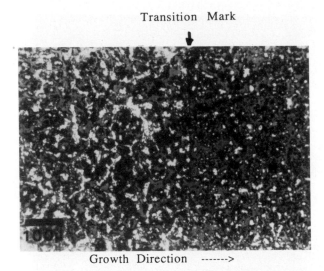

Growth Direction   ------->

Figure 2. **Microstructure of Al₂O₃/Al composites produced by regrowth at 1200°C and initiated using a sample from the bottom of the Al₂O₃/Al composite originally produced by direct oxidation of pure Al doped with MgO powder at 1200°C.**

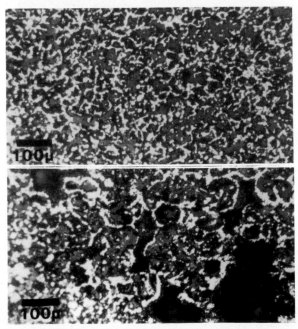

Figure 3. **Microstructures of Al₂O₃/Al composites (from near the metal/ ceramic interface) produced by direct oxidation of pure Al doped with 20 mg cm⁻² MgO powder at 1200°C. (a) 20 μm–32 μm. MgO particles, (b) 50 μm–100 μm MgO particles.**

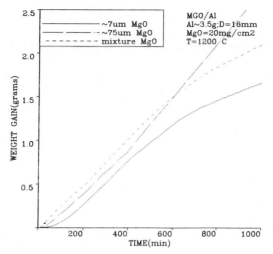

Figure 4. Weight gain versus reaction time at 1200°C for pure Al with the surface doped with 20 mg cm⁻² MgO of different particle sizes.

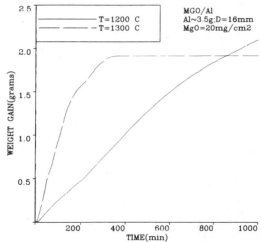

Figure 5. Weight gain versus reaction time for pure Al with the surface doped with 20 mg cm⁻² MgO at 1200°C and 1300°C respectively.

although there is still Al left below the composite after about 6 hours. Figure 6 shows the reaction rate per minute at 1200°C and 1300°C. There is considerable variation in reaction rate with time over short periods.

Microanalysis was carried out to determine elemental distributions in the MgO doped samples. For the DIMOX of MgO/Al and Al-Mg-Zn, the specimens were analysed before the Al was exhausted [6, 7, 8]. A high Mg concentration was always discovered at the top surface. This is expected to be in oxide form, either as MgO or MgAl$_2$O$_4$. Figure 7(a) shows a line microanalysis scan across the top surface (air/oxide interface) of a material grown for 5 hours at 1200°C with 17 mg cm^{-2} MgO surface dopant. Figure 7(b) shows the top surface of the material regrown from the bottom piece of MgO/Al DIMOX product, (the microstructure from the middle of the product is shown in Figure 2) and there is still a high MgO concentration at the top surface. However, the product of MgO/Al reaction at 1300°C for 16·6 hours still shows some Al left beneath the Al$_2$O$_3$/Al composite and there is no high MgO concentration at the surface (Figure 7(c)).

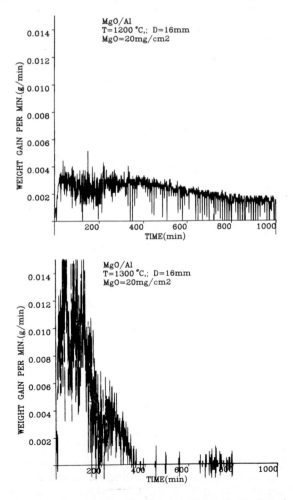

Figure 6. Weight gain rate versus reaction time for pure Al doped with 20 mg cm^{-2} MgO showing a considerable variation in reaction rate with time over short periods. (a) at 1200°C, (b) at 1300°C.

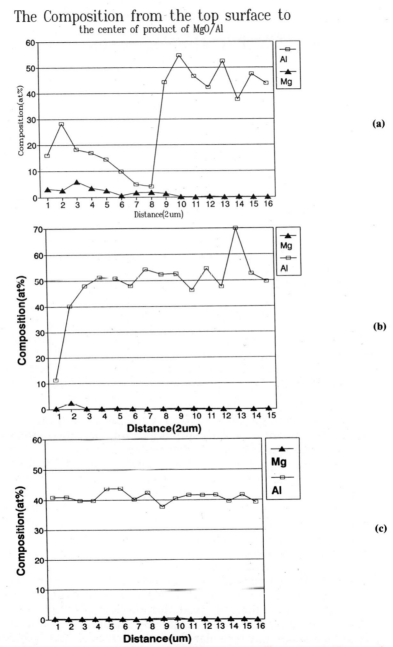

Figure 7. Microanalysis line scans across the air/ceramic interface of the product, (a) 27 mg cm^{-2} MgO doped Al, oxidised at 1200°C showing surface Mg segregation and (b) regrowth at 1200°C using a piece of the Al$_2$O$_3$/Al composite (see Figure 2), (c) 20 mg cm^{-2} MgO doped Al, oxidised at 1300°C.

5. DISCUSSION

A number of mechanisms have been proposed to explain the direct oxidation of the Al-Mg system [1, 8, 10, 6]. Detailed discussion of these different mechanisms has been reported elsewhere [8]. We have proposed a sequence of events which could lead to rapid oxide growth in Al-Mg systems. Consider a pure Al melt with a surface doping of dispersed MgO particles. When the Al is exposed to oxygen in the atmosphere the reaction:

$$2/3Al + 1/2O_2 = 1/3Al_2O_3 \tag{1}$$

occurs. At certain regions on the surface there will be MgO particles present and here an interdiffusion reaction is possible to form spinel:

$$Al_2O_3 + MgO = MgAl_2O_4 \tag{2}$$

In the case of magnesium aluminium spinel the reaction proceeds by the diffusion of both Mg + + and Al + + + ions which can lead to a possible disruption of the Al_2O_3 surface film. Breaks in the film allow the liquid metal to flow and wet the spinel or any MgO which has not reacted with Al_2O_3. In the case of pure Al this wetting will dissolve Mg from spinel or MgO until the phase equilibrium between Al_2O_3, $MgAl_2O_4$ or MgO, and Al (containing some Mg) is achieved, these reactions being given by:

$$MgAl_2O_4 + 2/3Al = 4/3Al_2O_3 + Mg \tag{3}$$

$$3MgO + 2Al = Al_2O_3 + 3Mg \tag{4}$$

However, the presence of the free surface will perturb this equilibrium because the high vapour pressure of Mg allows some evaporation. Thus, rather than returning to solution some Mg is lost to the atmosphere. However, once in vapour form the Mg can now react with oxygen to form MgO:

$$Mg(g) + 1/2O_2(g) = MgO(s) \tag{5}$$

Some will return to the surface ensuring the presence of MgO, disrupting any stable oxide film and allowing continuous rapid oxidation. Some Mg is lost to the environment, especially at higher temperatures.

The proposed mechanism has four stages: (1) a rapid surface oxidation of Al to Al_2O_3, (2) an interdiffusion stage which ruptures the oxide film, (3) a rapid wetting of $MgAl_2O_4$ or MgO leading to a further reaction and the partial loss of Mg to the atmosphere, (4) a deposition of MgO on the surface by vapour phase precipitation. The four stages recycle. The channel thickness of Al in the composite is determined by the rupture of the Al_2O_3 layer, *i.e.* Reaction 2. Initially the Al_2O_3 layer reacts with MgO powder on the surface. The channel distribution of Al in the early reaction product is thus determined by the size of MgO powder. The channels of Al in the later reaction product will be determined by the size of MgO from Reaction 5 (Figure 1). The channel thickness in the early oxidation product is found to be different for different sizes of MgO powder (Figure 3). Figure 2 also confirms this point. The initial reaction rate is similar if the top surface of Al is fully covered. After the recycling of MgO (Reaction 5), the reaction rates are different for different sizes of MgO (Figure 4). At high temperature MgO is more unstable compared with Al_2O_3, since $\Delta H_f MgO = -601 \cdot 6$ KJ/mol, whereas $\Delta H_f 1/_3 Al_2O_3 = -559 \cdot 6$ KJ/mol.

The Mg vapour pressure is high and Mg easily evaporates. At 1300°C the oxidation stops before all the Al is exhausted because Mg is lost and there is no MgO on the top surface (Figure 5). Figure 7 also confirms this point. We also analysed Mg concentrations at different positions in the product. The average Mg concentration is about half of the calculated concentration based on the amount of MgO powder used and assuming that there was no loss of Mg during oxidation.

6. CONCLUSIONS

The presence of MgO at the top surface is necessary to initiate direct oxidation. The size of MgO powder used affects the microstructure of the early oxidation product, but doesn't affect the early oxidation rate. The mechanism we propose is consistent with the experimental results observed.

REFERENCES

1. NEWKIRK, M. S., URQUHART, A. W., ZWICKER, H. R. & BREVA, E., *J. Mater. Res.,* **1**, 81, (1986).
2. AGHAJANIAN, M. K., MACMILLAN, N. H., KENNEDY, C. R. & LUSZCZ, J., *J. Mater. Sci.,* **24**, 658, (1989).
3. NAGELBERG, A. S., to be published in *J. Mater. Res.*
4. NAGELBERG, A. S., *Solid State Ionics,* **32/33**, 783, (1989).
5. NAGELBERG, A. S., *Proc. MRS Symp.,* **155**, 275, (1989).
6. XIAO, P. & DERBY, B., *Proc. Brit. Ceram. Soc.,* **48**, (1991).
7. XIAO, P. & DERBY, B., presented at 2nd Eur. Ceram. Soc. Conv., Augsburg, (1991), in press.
8. XIAO, P. & DERBY, B., submitted to *J. Eur. Ceram. Soc.*
9. NAGELBERG, A. S., ANTOLIN, S. & URQUHART, A. W., submitted to *J. Amer. Ceram. Soc.*
10. BAUM, L., SHAFRY, N. & BRANDON, D. G., presented at CIMTEC, (1990).

Oxidation of Silicon Powder: its Significance for the Strength of Reaction Bonded Silicon Nitride

R. G. STEPHEN and F. L. RILEY

School of Materials, University of Leeds, Leeds, LS2 9JT, U.K.

ABSTRACT

Silicon powder is readily oxidised by water of pH > 7 with the formation of thin surface coatings of silicon dioxide. At high pH the rate of reaction is fast: for example gentle agitation at 21°C with grinding media in water of pH 10 for 14 ks is sufficient to convert ~ 7% of ~ 8 μm silicon powder to silicon dioxide. Agitation in propan-2-ol (IPA) has no detectable effect. The coating consists of ~ 50 nm silicon dioxide particles which interact in the dried powder to give strong interparticle bonding. Dry powder consequently tends to agglomerate strongly, and marked microstructural inhomogeneity develops within compacted powder. Inhomogeneity is retained and is readily detected in the fully nitrided reaction bonded silicon nitride. The biaxial strength of compacted ~ 8 μm silicon powder increase from 2 MPa to ~ 5 MPa after agitation in water of pH 10, despite the presence of large agglomerates and correspondingly large voids. The strength of reaction bonded silicon nitride produced from these powders depends on the maximum void size. Nitridation of unoxidised silicon powder gave material with a maximum void size typically of ~ 110 μm, and a mean biaxial strength of ~ 290 MPa. Reaction bonded silicon nitride prepared from oxidised silicon contained a maximum void size of ~ 350 μm, and had strengths of ~ 190 MPa.

1. INTRODUCTION

The low temperature reaction between silicon powder and water:

$$Si_{(s)} + 2H_2O_{(l)} = SiO_{2(aq)} + 2H_{2(g)}; \quad \Delta G^{\circ}_{298} = -382 \text{ kJ mol}^{-1} \qquad (1)$$

is potentially important in the milling of silicon powder: [1–5]. Hydrogen evolution, which may be vigorous, can also occur during slip casting of silicon powder [6–10]. The rate of the oxidation reaction increases markedly with pH [11]. The silicon dioxide formed during actual milling, or by merely gentle agitation of the powder with milling media, is deposited as a strongly adhering coating of ~ 50 μm particles on the silicon particles [12]. Dry silicon dioxide coated silicon powder compacts under low pressing pressure (< ~ 150 MPa) to a lower density than an unoxidised control powder which had been agitated under otherwise identical conditions in propan-2-ol (IPA). This reflects the low bulk density of the coating phase. At high compaction pressures (> ~ 150 MPa), oxidised silicon powder gives densities much higher than those obtained from unoxidised powder.

This programme extended earlier work to assess the influence of such silicon dioxide coatings on the microstructure and strength of compacted silicon powder, and of the derived reaction bonded silicon nitride. The influence of the silicon dioxide coating on nitridation kinetics was also investigated.

2. EXPERIMENTAL

The silicon powder was KemaNord Sicomill IIC of median equivalent spherical diameter 8 μm ("Sedigraph 5000 ET," Micromeritics) 25 g batches of powder were mixed with 75 cm³ of IPA, or water of pH 5, 7, 8, 9 and 10 obtained using aqueous hydrochloric acid or ammonium hydroxide, and gently agitated by the vibratory action of a laboratory flask shaker for ~ 7 ks using ~ 450 g of stabilised zirconia media in a 500 cm³ polypropene flask. Hydrogen evolved was collected over water, and the amount of silicon dioxide produced was calculated from the total volume of gas evolved on the basis of Equation (1). On a larger scale, for preparative work, 50 g batches of silicon powder were agitated for 14 ks in IPA ("silicon IPA") and water of pH 5 ("silicon pH 5") and 10 ("Silicon pH 10") using ~ 900 g of stabilised zirconia media in a 1000 cm³ polypropene flask. Powders were dried at ~ 90°C using rotary vacuum evaporation, and passed through a 250 μm mesh nylon sieve. Discs of ~ 1 g of silicon powders IPA, pH 5 and pH 10 were uniaxially pressed in a steel die at 25 MPa and then iso-pressed at ~ 225–250 MPa to a constant density of 1·40 Mg m⁻³. The diameter and thickness of iso-pressed discs were ~ 17 mm and 3·0 mm respectively. Biaxial disc flecture strengths of compacted, and of nitrided, discs were measured using a standard testing machine ('Instron 1185,' High Wycombe, U.K.) at a cross-head speed of 50 μm min⁻¹, with concentric rings of hardened steel balls, of 16 and 4 mm diameter. Strength (S) was calculated using the expression [13]*

$$S = \frac{3W}{2\pi t^2} \left\{ (1 - v)\ln \frac{r_S}{r_L} + \frac{[(1 - v)(r_S^2 - r_L^2)]}{2r_D^2} \right\} \tag{2}$$

where
W = load at failure,
t = disc thickness,
v = Poisson's ratio of the disc material,
r_L, r_S and r_D = radii of the load multiball ring, support ball ring and the disc.

Isolated particles of silicon IPA and silicon pH 10 were examined by transmission electron microscopy (TEM). A small quantity of powder was dispersed in acetone, and a drop evaporated on a standard copper grid which had previously been carbon coated to support the particles and ensure good electrical conductivity.

Six discs of each batch of powder were nitrided using flowing $N_2/5\%$ H_2, in a vertical tube furnace with continuous measurement of weight changes (Mk. 3 Microforce balance, CI Electronics, Salisbury, U.K.). Samples were lowered over 5 s into the pre-equilibrated furnace hot zone, under the nitriding atmosphere. When the rate of weight increase at 1370°C had become negligible the temperature was increased to 1410°C, and subsequently to 1450°C in order to achieve complete nitridation.

---

*Reference 13 contains a typographical error in this expression.

Samples of compacted silicon powder, and reaction bonded silicon nitride, were impregnated with methyl methacrylate monomer which was then polymerised. These discs were mounted in Araldite resin and polished to 1 μm for microstructural examination by light microscopy (LM). The maximum void diameters (D_{max}) in polished cross-sections were measured using high contrast photographs.

3. RESULTS

The yield (mass%) of silicon dioxide after ~ 7 ks agitation at 21°C as a function of suspending fluid and its pH increases rapidly at pH > 8 (Figure 1).

TEM micrographs of silicon particles agitated for 7 ks in IPA, and in water at pH 10, are shown in Figures 2(a) and 2(b). The ~ 50 nm silicon dioxide spherical nodules forming the coating can clearly be seen (Figure 2(b)). There is an approximately linear relationship between strength of the compacted silicon powder discs and the yield of silicon dioxide (Figure 3). LM micrographs of compacted silicons IPA, pH 5 and pH 10 (Figures 4(a), 4(b) and 4(c)) show a general increase in microstructural inhomogeneity, and in the largest void size in particular, with increasing silicon dioxide content.

Initial rates of silicon nitride formation at 1370°C were very slow for extensively water-oxidised silicon powder. The nitridation rate of silicon pH 10 accelerated after 1 ks but after 2 ks the silicon was only 32% nitrided. There was a much shorter induction period of silicon pH 5, but the extent of nitridation after the first 2 ks was still less than that of silicon IPA (Figure 5). Nitridation rates for silicon IPA and pH 5 at 1370°C reached a maximum of 3–3·5 mg s⁻¹; the peak nitridation rate of silicon IPA was reached slightly before that of silicon pH 5 (Figure 6).

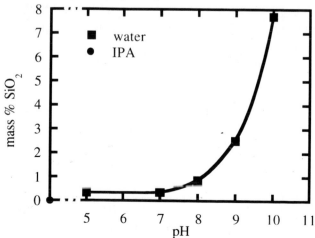

Figure 1. Silicon dioxide yield after 7 ks of agitation, as a function of the type and pH of the suspending fluid.

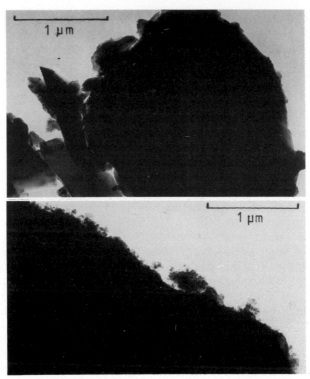

Figure 2(a) and 2(b). TEM micrographs of silicon particles filled in IPA and water of pH 10.

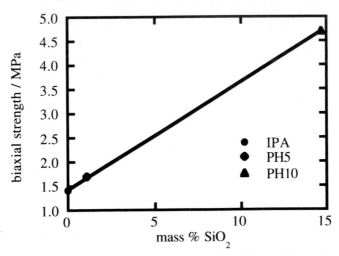

Figure 3. Biaxial strength of compacted silicon agitated for 14 ks, in IPA and in water of pH 5 and 10, as a function of silicon dioxide yield.

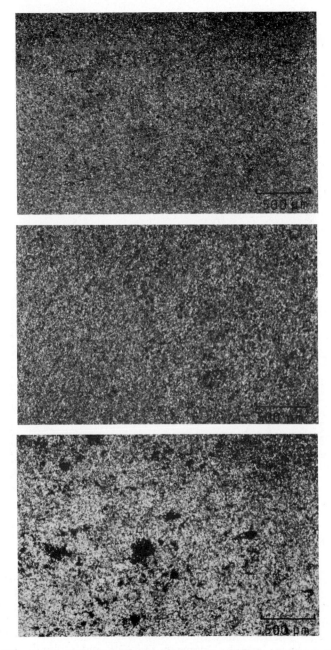

Figure 4(a), 4(b) and 4(c). LM micrographs of polished sections of compacted silicon samples IPA, pH 5 and pH 10.

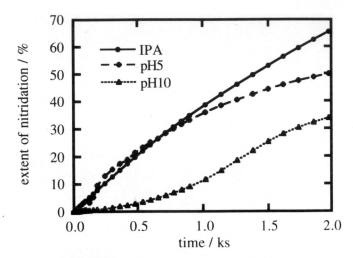

Figure 5. Extent of nitridation at 1370°C as a function of time, for silicon powders agitated for 14 ks in IPA and water of pH 5 and 10.

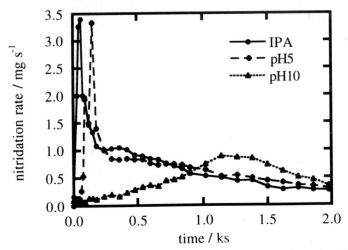

Figure 6. Nitridation rate at 1370°C as a function of time for silicon powders agitated for 14 ks in IPA and water of pH 5 and 10.

In contrast to the effect seen with compacted, unnitrided silicon powder (Figure 3), an increase in the amount of silicon dioxide coating was associated with a decrease in the biaxial strength of nitrided discs (Figure 7). Highest strengths of reaction bonded silicon nitride were obtained for materials containing the smallest maximum void size (D_{max}), and there was an approximately linear relation between $D_{max}^{-1/2}$ and strength, as shown in Figure 8. LM micrographs of the reaction bonded silicon nitride (Figures 9(a),

9(b) and 9(c)) show that the pore size distribution for nitrided silicon IPA was the narrowest, and that considerable inhomogeneity occurred in the structure of nitrided silicon pH 10. This material also contained small particles of unreacted silicon.

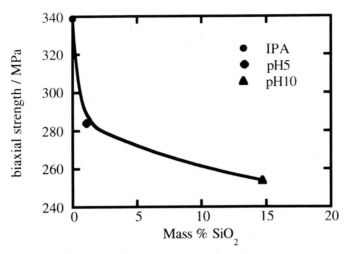

Figure 7. Biaxial strength of nitrided discs prepared from silicon agitated for 14 ks in IPA and water of pH 5 and 10, as a function of silicon dioxide yield.

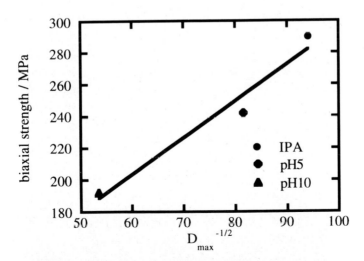

Figure 8. Biaxial strength of nitrided discs prepared from silicon agitated for 14 ks, in IPA and water of pH 5 and 10, as a function of $D_{max}^{-1/2}$.

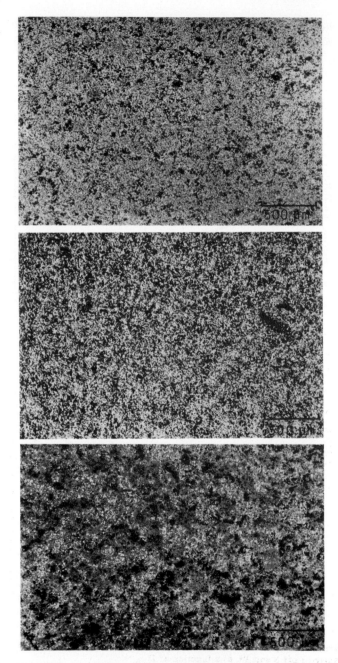

Figure 9. LM micrographs of polished sections of silicon (a) IPA, (b) pH 5 and (c) pH 10 initially nitrided at 1370°C, and finally at 1450°C.

4. DISCUSSION

The gentle agitation of silicon powder in water of $>$ pH 5 produces in a few minutes readily measurable quantities of hydrogen and an equivalent amount of silicon dioxide; at pH 10 a large amount of amorphous silicon dioxide is rapidly produced, the ultimate form of which is the 50 nm nodular coating on the silicon particles clearly shown by the TEM photographs. It appears from its location and morphology that the coating is nucleated and grown during oxidation, and not subsequently during the drying stage, and for which there is independent evidence [14]. There is no sign that significant fracture of the silicon particle occurs during this agitation process. The nm size silicon dioxide particles at the silicon particle surfaces provide a very low density cushioning film which leads to an overall low bulk density for the silicon powder on compaction at low pressures; they also serve to bond strongly the silicon particles and enhance markedly the natural tendency for agglomerate formation to occur. These agglomerates appear to be quite hard, and visibly (fracture surface SEM photographs) survive compaction pressure of up to 250 MPa [15]. Such strong bonding between compacted nm size particles also gives rise to the very high green strengths as shown by Figure 3, and observed also in other similarly coated powder systems (alumina/zirconia [16, 17], and silicon nitride/alumina [18]). Kendall et al. have provided a fracture mechanics model for the green strength of a compacted powder, which predicts strengths similar to those seen here assuming critical defect sizes in the 100 μm to 300 μm range [19]. In the silicon powders used in this work the increasing strength of the interparticle bonding of the oxide-coating with increasing silicon dioxide content appears to more than compensate for the larger void sizes seen in the agglomerated and inhomogeneously compacted powder. Similar coating volume-green strength relationships have been observed with other systems [16, 17, 18] although in those cases of finer ($\sim$ 500 nm) primary particles, packing inhomogeneity in the green state was not detected.

The inhomogeneity of particle packing, and specifically the void size distribution, is not changed significantly on nitridation. In the resulting reaction bonded silicon nitride the strengthening effect of the nm size silicon dioxide particles seen in the precursor green powder is lost because of the volatilisation of the silicon dioxide as SiO, with the subsequent formation of silicon oxynitride. The void (critical defect) size effect now dominates, and the measured strength appears to follow at least approximately the Griffith relationship:

$$S = \frac{Z}{Y} \frac{(2E\gamma)}{D_{max}}^{\frac{1}{2}} \qquad (3)$$

where S is the tensile strength, D_{max} is the largest defect (or void) dimension, Z is a flaw shape parameter ($\sim$ 1·6 for circular flaws), Y is a loading and crack configuration parameter, and E and γ have their usual meanings [20]. Inserting measured values of D_{max} and S into this expression and setting

$Z/Y = 1$ gives values of K_{Ic} of the order of 3 MPa m$^{1/2}$, which is within the range of values reported for commercial reaction bonded silicon nitride materials of densities similar to that obtained here (2·40 Mg m^{-3}) [21] Any factor influencing the packing homogeneity of the green silicon powder must have a strong influence on the strength of the reaction bonded silicon nitride, and it is clear that the surface films of nm dimension silicon dioxide particles are very important in this context.

A second important function of the oxide film is to control the initial nitridation rate. This action of silicon dioxide is well documented [22, 23] and its effect in the case of the deliberately over-oxidised silicon pH 10 powder is striking. There are lesser but still clearly detectable differences between the initial nitridation rates of silicon pH 5 and silicon IPA powders, which shows that significant oxidation of silicon has occurred in water of pH 5, even though the volume of hydrogen evolved during agitation was at the limits of detection.

It must be concluded that any low temperature liquid phase oxidation of silicon, which forms particulate silicon dioxide (in contrast to the dense, thin films of silicon dioxide formed during high temperature oxidation) is likely to be detrimental to the packing characteristics of the powder, if a vigorous effort is not made to prevent the formation of, or to break down, agglomerates before compaction. From the results of this study there was no doubt that the best quality reaction bonded silicon nitride was obtained through the use of the non-aqueous milling fluid.

5. CONCLUSIONS

Gentle agitation of silicon powder with grinding media in aqueous solutions of pH $\geqslant$ 5 leads to the rapid formation of hydrogen and silicon dioxide. At pH 10 oxidation is very fast and the silicon dioxide is deposited as a 50 nm nodular film on the silicon particle surfaces. These oxide films cause strong agglomeration in the dry powder, and subsequently the development of marked inhomogeneity, particularly with regard to the distribution of void size, in compacted powder. The green strength of compacted silicon powder is directly proportional to the amount of silicon dioxide produced during oxidation. Microstructural inhomogeneity in the silicon powder is carried forward on nitridation into the reaction bonded silicon nitride, with a consequent detrimental effect on strength. There is a strong correlation between the maximum observed void size in reaction bonded silicon nitride, and its measured strength. The highest strength reaction bonded silicon nitride was obtained from silicon powder which had been processed in IPA.

It is clear that the state of oxidation of a silicon powder is an important factor influencing all stages of the processing of the powder to reaction bonded silicon nitride, and ultimately controlling the strength of this material.

ACKNOWLEDGMENTS

This work was supported by a SERC CASE Studentship in associated with T. & N. Technology Ltd., Rugby, U.K. The assistance of N. Han with transmission electron microscopy is acknowledged.

REFERENCES

1. KOLBANEV, I. V. & BUTYAGIN, P. YU., *Kinet. Katal.,* **23**, 327, (1982).
2. MONTENYOHL, V. I. & OLSON, C. M., U.S. Patent No. 2 614 993, (1952).
3. KEMPER, C. P. & ALVEREZ-TOSTADO, C., U.S. Patent No. 2 614 883, (1952).
4. ISMAIL, Z. K., HAGUE, R. H., FREDIN, L., KAUFMANN, J. W. & MARGRAVE, J. L., *J. Chem. Phys.,* **77**, 1617, (1982).
5. TACHIVANA, A., KOIZUMI, M., TERAMANE, H. & YAMABE, T., *J. Am. Cer. Soc.,* **109**, 1387, (1987).
6. EZIS, A., in *Ceramics for High Performance Applications,* Eds. J. J. Burke, A. E. Gorum and R. N. Katz, Brook Hill Publishing, Chestnut Hill, MA, 207, (1974).
7. SACKS, M. D. & SCHEIFFELE, G. W., *Ceram. Eng. Sci. Proc.,* **6**, 1109, (1985).
8. NAWANO, K. & MOULSON, A. J., Unpublished work, University of Leeds, U.K.
9. WILLIAMS, R. M. & EZIS, A., *Am. Ceram. Soc. Bull.,* **62**, 607, (1984).
10. SACKS, M. D., *Am. Ceram. Soc. Bull.,* **63**, 1510, (1984).
11. STEPHEN, R. G. & RILEY, F. L., *J. Eur. Ceram. Soc.,* **5**, 219, (1989).
12. STEPHEN, R. G. & RILEY, F. L., *J. Eur. Ceram. Soc.,* **9**, (4), 301–307, (1992).
13. GODFREY, D. J., *Brit. Ceram. Proc.,* **39**, 133, (1987).
14. STEPHEN, R. G. & RILEY, F. L., unpublished work.
15. STEPHEN, R. G. & RILEY, F. L., *Proc. 4th Int. Symp. on Ceramic Materials and Components for Engines,* Eds. R. Carlsson, T, Johansson and L. Kahlman, Elsevier Applied Science (London), pp. 307–314, (1992).
16. BROMLEY, A. P., RILEY, F. L., DRANSFIELD, G. P. & EGERTON, T. A., *Solid State Phenomena,* Vol. 25 & 26, pp. 97–104, (1992).
17. BROMLEY, A. P., RILEY, F. L., DRANSFIELD, G. P. & EGERTON, T. A., to be published in *Proc. 2nd Eur. Ceram. Soc. Conf.,* September 11–14, Augsburg, F.R.G., (1991).
18. WANG, C.-M. & RILEY, F. L., to be published in *Proc. 2nd Eur. Ceram. Soc. Conf.,* September 11–14, Augsburg, F.R.G., (1991).
19. KENDALL, K., ALFORD, M. McN. & BIRCHALL, J. D., in *Special Ceramics,* **8**, *Brit. Ceram. Proc.,* **37**, 255, (1986).
20. DAVIDGE, R. W., *Mechanical Behaviour of Ceramics,* Cambridge University Press, Cambridge, (1979).
21. RILEY, F. L., in *Preparation and Properties of Silicon Nitride Based Materials,* Eds. D. A. Bonnell and T. Y. Tien, *Materials Science Forum,* **47**, Trans Tech Publications, Aedermannsdorf, 70, (1989).
22. CAMPOS-LORIZ, D. & RILEY, F. L., *J. Mat. Sci. Lett.,* **11**, 195, (1976).
23. DERVISBEGOVIC, H. & RILEY, F. L., *J. Mat. Sci.,* **13**, 1945, (1978).

Fabrication of Nitrogen Ceramics with Low Temperature-Coefficients of Permittivity

A. M. DICKINS, J. S. THORP* and D. P. THOMPSON

Wolfson Laboratory, Materials Division, Dept. of Mechanical, Materials & Manufacturing Engineering, University of Newcastle upon Tyne, NE1 7RU
*School of Engineering and Applied Science, University of Durham, Durham, DH1 3LE

ABSTRACT

Ceramics based on Si_3N_4 or AlN are potentially attractive materials for radomes, but this application requires a temperature-stable permittivity and this is a condition which nitrogen ceramics have not yet been able to satisfy. In the present study, silicon nitride, β'-sialon and aluminium nitride have been investigated in an attempt to overcome this problem. The solution adopted is known as 'compensation' and involves adding a phase having a permittivity which decreases with temperature, so the overall permittivity of the composite ceramic is essentially temperature-stable.

Most paraelectrics are titanates or zirconates and cannot be used to compensate nitrogen ceramics because they react with the matrix at the temperatures required for densification. Although silicon nitride or β'-sialon would be preferred as the matrix phase because of their superior thermomechanical properties, no suitable compensator has yet been found for these materials. However, calcium hexaluminate ($CaAl_{12}O_{19}$) and strontium hexaluminate ($SrAl_{12}O_{19}$) are retained when sintered with certain sialon ceramics (AlN and the AlN-polytypoids: 8H, 15R, 12H, etc.), and several composite materials have been fabricated. Dielectric measurements at 9·375 GHz confirm that compensation does occur.

Gas Pressure Sintering of Silicon Nitride: The Influence of Additive Content and Sintering Parameters on Density

J. SLEURS, R. GILISSEN, R. C. PILLER* and R. W. DAVIDGE*

SCK/VITO, Boeretang 200, B-2400 Mol, Belgium
*AEA Technology, Harwell, OX11 0RA, UK

ABSTRACT
Two stage gas pressure sintering, a variant of Sinter/HIP, was applied to silicon nitride powder compacts containing varying levels of sintering aids. The experiments show that, if the appropriate conditions are satisfied, full densification is achieved. The benefits of the process are also discussed.

1. INTRODUCTION

The theoretical background to gas pressure sintering was developed by Prochazka and Greskovich [1]. Figure 1 shows the equilibrium diagram for N_2 and Si vapour pressures above Si_3N_4. The coexistence limit between silicon and Si_3N_4 runs from bottom left to top right. It represents nitrogen pressures above which Si_3N_4 exists as solid if there is an equilibrium partial pressure of silicon. This implies that Si_3N_4 will also decompose at high nitrogen pressures if silicon vapour is not prevented from escaping from the system and lowering the silicon equilibrium partial pressure. To avoid spontaneous decomposition of silicon nitride during sintering it is general practice to use 10 times higher nitrogen pressures than those corresponding to the coexistence limit (dashed line in Figure 1 — *e.g.* maximum 1800°C at 0·1 MPa N_2 pressure). Increasing the nitrogen pressure at constant temperature lowers the corresponding Si-equilibrium vapour pressure so that the need for Si vapour pressure build up becomes less critical. On the other hand, a higher nitrogen pressure permits sintering at higher temperatures (*e.g.* 2200°C at 10 MPa N_2 pressure).

The simultaneous increase of temperature and N_2-pressure make it possible to obtain higher densities after sintering [1]. However, full densification can never be achieved by one step pressure sintering. Densification is hindered by the inclusion of gas under pressure in the porosity [2]. The use of higher sintering temperatures also decreases the fracture toughness due to grain coarsening and globularisation of the rod-like β-Si_3N_4 grains [3].

Attempts to overcome these drawbacks have led to the development of a two-step sintering cycle [1, 4]. In the first step the material is sintered at normal or slight overpressure until a closed porosity stage is reached. In the second step the densification is continued under the combined effect of temperature and increased pressure. Two-step gas pressure sintering is in fact a sinter-HIP process at moderate pressure. If sintering conditions are found which lead in the first step to closed porosities, full density can be achieved at pressures less or equal to 10 MPa. The process can also heal microstructural defects.

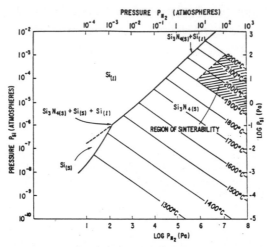

Figure 1. N_2 and Si equilibrium pressures above Si_3N_4 [1].

Artificially introduced pores of about 50 μm are closed by two-step gas pressure sintering [3]. Fracture toughness can be significantly increased up to an optimum temperature above which K_{IC} decreases by grain coarsening [3]. The bend strength of high density pressureless sintered materials can also be increased by two-step gas pressure sintering. This can be explained by the elimination or reduction of large process related microdefects (*e.g.* consolidation faults and dust inclusions) which determine fracture strength.

The influence of different material parameters on the densification behaviour of Si_3N_4 with Y_2O_3 as a sintering additive is summarised in two recent articles [2, 5]. A literature survey on gas pressure sintering has been published by Kolaska [6].

2. EXPERIMENTAL

All experiments have been executed with UBE Si_3N_4 SN-E10, the characteristics of which are summarised in Table 1. Three sets of powder mixtures have been compared with a pressureless sinterable standard mixture. The

Table 1. Characteristics of UBE SN-E10
Si_3N_4 powder

Specific surface area m^2/g	13·7
Mean particle size d_{50} (μm)	0·35
Powder phase: α-content (%)	97
Chemical composition (wt%)	
O	1·01
C	0·08
Metallic impurities (ppm)	
Fe	43
Al	< 30
Ca	15

Table 2. Composition of Si$_3$N$_4$ powder mixtures

	Si$_3$N$_4$	Y$_2$O$_3$	Al$_2$O$_3$
Reference mixture			
SN 446-1	93	5	2
Mixtures with decreasing sintering additive content			
SN 442	96·5	2·5	1
SN 443	98·5	1·25	0·5
Mixtures without Al$_2$O$_3$			
SN 444-2	95	5	—
SN 445	97·5	2·5	—

composition of the different mixtures are summarised in Table 2. Powders are mixed by planetary ball milling with sintered Si$_3$N$_4$ balls in polyethylene jars and isopropanol as milling fluid.

The mixtures are pelletised at 1·5 t/cm^2 into cylindrical pellets of 10 mm diameter and approximately 10 mm height. Green density is about 1·60 g/cm^3 (50% of theoretical density).

The following sintering cycles have been performed:

+ 10°C/min, 1 h at temperatures up to 1800°C, 0·1 MPa N$_2$, pressure;

+ 10°C/min, 1 h at temperatures up to 1850°C, 0·5 MPa N$_2$, pressure;

+ 10°C/min, 1 h, 0·1 or 0·5 MPa N$_2$ pressure followed by a second step at 1780°C, 10 MPa N$_2$ pressure for 1 h.

All density measurements are made by the water immersion technique and are reported as % relative density, taking the theoretical density of pure Si$_3$N$_4$ (3·20 g/cm^3) as the reference density. The actual density is somewhat higher due to the presence of different yttrium-containing, amorphous and crystalline grain boundary phases but cannot be exactly taken into account. This explains why densities above 100% are reported.

Microstructure has been examined by optical microscopy on polished sections. Very small pores are difficult to visualise as silicon nitride is translucent and high magnifications are not practicable.

3. RESULTS AND DISCUSSION

The results of two step gas pressure sintering on the different powder mixtures under investigation are shown in Figure 2. The figure gives the relation between the density obtained after the first step and the density after the full cycle. It can be seen that, if after the first step a density of more than 92% is reached, full densification can be obtained during the gas pressure step. This is in agreement with the generally accepted idea that no open porosity persists at densities above 92–93%. The increase of approximately 2% densification observed in materials with densities below 90% is only due to the longer heat treatment, the only positive effect of the nitrogen pressure being the suppression of weight loss during the second hour of sintering.

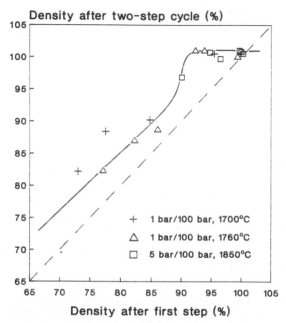

Figure 2. **Two-step gas pressure sintering: relation between densities after the first step and densities after completion of the full cycle.**

In order to define conditions to achieve closed porosity the first step of sintering has been studied in more detail for the different mixtures. Densities obtained after 1 h sintering under 0·1 and 0·5 MPa of nitrogen pressure are summarised in Figures 3 and 4.

The reference powder mixtures (SN 446-1) readily sinters under 0·1 MPa nitrogen pressure to nearly full density at 1800°C. The microstructure shows homogeneously distributed fine pores.

Mixtures containing less sintering additives are attractive because they will give better high temperature strength and better oxidation resistance where they can be sintered to high density. These mixtures cannot be sintered to closed porosity at 0·1 MPa N_2 pressure. For samples containing only Y_2O_3 the densities even drop at the highest practicable sintering temperature of 1800°C (Figure 3). This can probably be explained by the important weight loss throughout the open pore structure which persists during the total sintering cycle. Sintering at 1850°C under 0·5 MPa yields densities above 95% for two mixtures. No higher temperatures have been used, to protect the thermocouple assemblages of the HIP. From Figure 1 it can be seen that temperatures can be increased without harm beyond 1900°C. Extrapolating the density curves of Figures 3 and 4 gives densities above 95% even for the lowest sintering additive concentrations (1·25% Y_2O_3 + 0·5 Al_2O_3 or 2·5% Y_2O_3 without Al_2O_3).

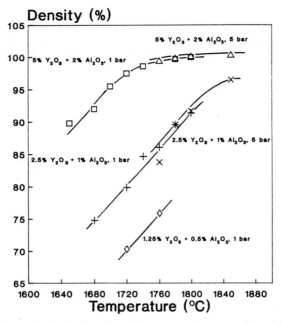

Figure 3. Effect of sintering additive (5:1 — Y₂O₃:Al₂O₃) concentration on density after sintering.

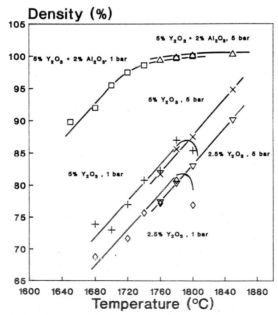

Figure 4. Effect of sintering additive (including Y₂O₃ alone) concentration on density after sintering.

The breakdown of hard agglomerates often encountered in fine sinterable Si_3N_4 starting powders and the homogeneous dispersion of sintering additives throughout the powder mixture are prerequisites for obtaining good and reproducible material characteristics [6, 7]. It is common practice to wet-mill the powder mixture to obtain this goal.

Intensive milling has two side effects: increasing the specific surface area and increasing the oxygen content. Both will increase sinterability but a higher oxygen content will have an adverse effect on high temperature properties. To determine if powder processing is equally important when applying two step gas pressure sintering less well processed powders are included in the study. Inadequately processed mixtures do sinter to densities above 95% under 0·1 MPa N_2 pressure at 1800°C. Under 0·5 MPa N_2 pressure at 1850°C even densities near 100% are obtained. Micrographs reveal however the existence of large pores which will affect fracture strength.

Ceramography shows that in all cases where the closed porosity stage is reached in the first step full densification occurs during the second step. In the case of badly processed mixtures some residual large pores can be seen.

4. CONCLUSIONS

Two-stage gas pressure sintering can produce full densification regardless of the concentration of additives, the degree of processing and the sintering conditions during stage 1 if the closed porosity condition ($> 92\%$) is reached during the first stage. The benefit of this process is that lower sintering temperatures can be applied during the first stage, thus reducing weight loss and suppressing the formation of near surface layers. The possibility of densifying mixtures containing less additives should lead to better high temperature properties.

ACKNOWLEDGMENTS

This research has been carried out jointly by SCK-Mol (B), AEA Harwell (UK) and NFE/IMT St. Niklaas (B) within the CEC Euram Programme. Partial financial support was obtained from CEC (contract MA1E.0022C) and from Céramétal (L), BP International Ltd., RTZ Chemicals, ICI plc., RAE and T. & N. Technology Ltd. (all UK).

The authors gratefully acknowledge this support and they wish to thank their colleagues and collaborators for carefully executing the experimental work and for their assistance.

REFERENCES

1. GRESKOVICH, C., in 2nd NATO ASI: *Progress in Nitrogen Ceramics,* Ed. F. L. Riley, publ. Nyhoff, Boston, 283, (1983).
2. WÖTTING, G. & ZIEGLER, G., *Ceramic Forum International,* **65**, 364, (1988).
3. ZIEGLER, G. & WÖTTING, G., in *Sinter/HIP-Technologie,* Ed. H. Kolaska, publ. Verlag Schmid GmbH, (1987).
4. KATZ, R. N., in 2nd NATO ASI: *Progress in Nitrogen Ceramics,* Ed. F. L. Riley, publ. Nyhoff, Boston, 3, (1983).
5. WÖTTING, G. & ZIEGLER, G., *Ceramic Forum International,* **65**, 471, (1988).
6. KOLASKA, H. & DREYER, K., *Sinter/HIP-Technologie,* Ed. H. Kolaska, publ. Verlag Schmid GmbH, 22, (1987).
7. GILISSEN, R., in *Euro Ceramics,* **1**, Eds. G. de With, R. A. Terpstra and R. Metselaar, publ. Elsevier, London, 1155, (1989).

Binder-Free Silicon Nitride — A Self-Strengthening Phenomenon

MICHAIL M. GASIK* and FRANK R. SALE

Manchester Materials Science Centre, University of Manchester and UMIST, Grosvenor Street, Manchester, M1 7HS

ABSTRACT

A thermal strengthening treatment (TST), which occurs on the low temperature heating of binder-free compacts of silicon nitride, is reported and discussed. Thermal analysis, dilatometry, SEM with EDX and FTIR have been used in this preliminary investigation of the phenomenon. The strengthening is shown to occur in separate stages which occur over clearly identifiable regions of temperature. The chemical reactions which have been shown to participate in the strengthening process are the loss of physically and chemically combined water and the decomposition of ammonium carbonate and various hydro-silicates.

It is shown possible to obtain a green density of 70–80% theoretical with uniaxial pressing at 200 MPa using binder-free mixes of β-Si$_3$N$_4$ (containing 10% α phase) with fine oxide additions of Al$_2$O$_3$, MgO and Y$_2$O$_3$. Compacts have bend strengths of 8–10 MPa after TST at 500°C and 30–34 MPa after TST at 900°C, which allow mechanical handling and machining.

1. INTRODUCTION

Engineering ceramics based on silicon nitride are employed in many areas because they have high strength, good thermal shock and oxidation resistance. Modern applications include high-performance engines, the space industry, and energy and environmental technology. However, the production of high quality Si$_3$N$_4$ parts is a complicated process with several expensive stages being required to obtain the necessary properties in the "green," "brown," and sintered states. In this context there is a major challenge to devise simple and effective processing techniques that allow the fabrication of silicon nitride ceramics into dense parts of high reliability.

This paper extends results presented previously on a promising manufacturing method which allow the production of near-net-shape silicon nitride parts by simple uniaxial powder pressing without any binder addition [1]. The uniaxial pressing of these binder-free mixes of β-Si$_3$N$_4$ (containing 10% α-Si$_3$N$_4$), with additions of fine Al$_2$O$_3$, MgO and Y$_2$O$_3$, have been shown to give 70–80% theoretical density when ultra-fine alumina (20–200 nm particle size) has been used together with a bi-modal size distribution of silicon nitride particles [1].

The present study has been performed, using medium temperature dilatometry (maximum temperature 1000°C), simultaneous thermal analysis (TG/DTA), and fast Fourier-transform infrared spectroscopy (FTIR), to

*Visiting British Council Research Scientist, Dnepropetrovsk Metallurgical Institute, Dnepropetrovsk, Ukrain/USSR, now with Institute of Materials Technology, Helsinki University of Technology, ESPOO, Finland.

determine the processes taking place during the thermal strengthening
treatment (TST) of such powder mixes.

2. EXPERIMENTAL PROCEDURE

2.1 Materials

The following powders have been employed in the study:– silicon nitride (95%
β-phase) produced by self-propagating high-temperature synthesis (SHS),
which has been attritor milled to give a surface area 7 m^2/g; silicon nitride
(98% α-phase) manufactured by plasma-chemical synthesis (PCS) of silicon
powder using direct nitriding to give a surface area of 25 m^2/g; and PCS
produced alumina, magnesia and yttria powders, all with surface areas of
25 m^2/g. In addition small amounts (< 1%) of ultra fine oxides obtained by
precipitation of hydroxides and controlled low temperature decomposition
were also added to the powder mixes.

In accordance with the results of the preliminary experimentation [1] the
ratio of PCS to SHS silicon nitride was selected as 1:9, (*i.e.* close to 10% for
the α/α + β ratio), in order to achieve maximum green density. The reference
designations of the samples (used henceforth in this paper) include an initial
letter of an oxide following a number which represents its weight%
concentration in the powder mix. For example, designation 10AY3 means SHS
silicon nitride (major) with 10 wt% PCS silicon nitride, 4 wt% alumina and
3 wt% yttria.

2.2 Thermal Analysis

Simultaneous thermal analysis (TG/DTA) has been carried out on "press-
ready" powder mixes of 10A4Y5M3, 10A3Y2M2 and 10A4YM1, at a heating
rate of 10°C/min up to 600°C using Seiko STA 320/200 equipment. A flow
rate of air of 25 ml/min was used for all STA experiments.

Two mixes of powder, selected on the basis of previous optimisation to be
10A3Y2M2 and 10A4Y2M1 (*i.e.* containing a total oxide addition of 7 wt%
but of differing Al$_2$O$_3$:MgO ratio), were studied dilatometrically by scanning
at rates of 1–5°C min^{-1} in static air at temperatures up to 1000°C. The green
specimens for these studies, which were of dimensions 5 × 5 × 3 mm, were
pressed uniaxially at 0·2 GPa. Mass change data were determined by weighing
samples, which were cut off from these pressed Modulus of Rupture (MOR)
bars used for dilatometry and mechanical testing, after heating to various
critical temperatures indicated by the dilatometric data.

2.3 Material Characterisation

Fourier transform infrared spectroscopy (FTIR) was used to study surface
chemical changes that occurred on heating powder specimens. This technique
has been shown to give important data concerning changes in surface structure
and composition of advanced engineering ceramics [2]. The infrared
spectrum was determined using a "Nicolet 5DX" spectrometer with fast
Fourier transform of the resulting interferogram. The powders under study
were mixed with pure KBr (1·64 wt%) and then pressed in a steel die at

750 MPa to give thin discs. The resultant transmittance curves were smoothed by 17th point digital interpolation with a spectra resolution of 4 cm⁻¹.

Preliminary observations of fracture surfaces of pellets of 10A4Y1M3 were made by scanning electron microscopy (Philips SEM 525) after heat treatment at 200°C and after pressure assisted sintering at 1750°C for 30 minutes, 0·8 MPa pressure of nitrogen using boron nitride embedding powder. Three point bend tests have been used to determine the strength of pressed powder specimens after heat treatment at 200°C, 500°C and 900°C.

3. RESULTS AND DISCUSSION

3.1 Thermal Analysis

Figure 1 shows STA data up to 600°C for the mix 10A3Y2M2, which is representative of all mixes investigated. It can be seen that there are two main stages of decomposition in this temperature range, at 50–120°C and 220–400°C respectively. The reaction at 50–120°C is clearly a sharp endothermic process whilst that commencing at 220°C is a broad exotherm which is possibly a doublet caused by overlapping peaks. The DTA data are substantiated by the TG data which show a mass loss of some 3 wt% associated with the low temperature reaction and a further 3·5 to 4 wt% loss (in two overlapping stages) which is associated with the higher temperature process.

Equivalent dilatometric data are shown in Figure 1 where two stages of shrinkage can be seen to be associated with the mass changes and thermal events at 50–120°C and 220–400°C. The first of these is only of the order of

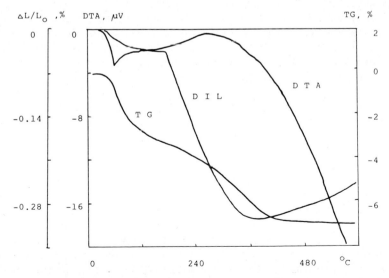

Figure 1. Thermal analysis of the 10A3Y2M2 powder mix (DTA/TG, 10°C min⁻¹, air flow 25 cm³ min⁻¹) and dilatometry (5°C min⁻¹, static air).

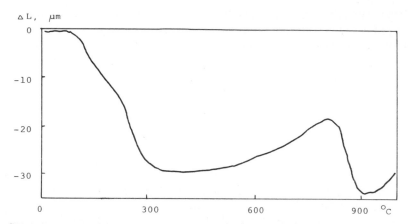

Figure 2. Dilatometric data for the 10AY4Y2M powder mix (1°C min⁻¹, static air).

0·04% whereas the second is more significant and is of the order of 0·3%. Figure 2 shows further dilatometric data, at a lower sensitivity but over the temperature range up to 1000°C. Here it is clear that there is a second significant contraction of the pellet which occurs over the temperature range 800–900°C. Between the two regions of significant shrinkage (220–400°C and 800–900°C) there is a steady expansion on heating.

The first slight contraction, which is accompanied by the sharp endothermic event and mass loss of the order of 3 wt% is associated with the loss of physically adsorbed water. However, during the powder processing prior to pressing there are many opportunities for the chemical association of water to occur by a range of hydrolysis reactions with the very fine powders used in this work. In particular the production of hydrated silica complexes on the surface of silicon nitride is of importance. Such reactions may be represented by:

$$[\text{Si}_3\text{N}_4] + (10 + 3y)\text{H}_2\text{O} \rightleftharpoons 3[\text{SiO}_2 \cdot y\text{H}_2\text{O}] + 4\text{NH}_4\text{OH} \qquad (1)$$
$$\text{surface} \qquad\qquad\qquad \text{Si}_3\text{N}_4 \text{ surface}$$

As a result ammonium hydroxide is produced as silica complex hydrates, with different 2D and 3D networks, are produced [3]. In the presence of other fine oxides, such as aluminium, these complexes may incorporate the other oxides as a result of the hydration process. This type knowledge of the surface chemistry of 'raw' and 'processed' powders and the production of such hydrated complexes is of vital importance in the successful utilization of fine powders and has received much attention recently [4, 5]. It is the decomposition of these complex hydrates that can be seen to occur over the temperature range 220–400°C to give the mass loss of approximately 3·5 wt% and the first significant shrinkage of the pellets. The products of decomposition over the range 220–400°C are very reactive, SiO₂-rich surface films. FTIR data presented in the next section help in the explanation of the low temperature processes. The low temperature results are also in agreement with Kawamoto *et al.* [6] who identified progressive loss of H₂O from silicon nitride by temperature-programmed desorption analysis (TPD).

The second major shrinkage, shown on Figure 2, which occurs over the temperature range 800-900°C and follows from the expansion on heating from 400°C-800°C, occurs whilst a mass gain results. The intermittent mass change measurements revealed that the maximum mass load of 6-7 wt% had occurred by 500-600°C. On subsequent heating from 600°C to 900°C a mass gain of 2-3 wt% results, which indicates that oxidation of the compacts has occurred. Microstructural examination of compacts previously heated to 800°C and to 900°C revealed that the fine Al_2O_3 particles located on the surfaces of the much larger Si_3N_4 particles, which were seen on all green samples and those heated up to 800°C, had disappeared on heating over the range 800-900°C. It is postulated that these fine particles reacted with the active decomposition products of the hydrated silica complexes and the freshly oxidised Si_3N_4 surfaces to produce glassy alumino-silicates as surface layers. The disappearance of the Al_2O_3 particles is clearly associated with the shrinkage over the range 800-900°C and with the development of increased strengths in the compacts. It seems, therefore, that the reaction product acts as an inorganic binder phase for the compacts. Fracture surfaces of compacts heated to temperatures of 850°C and above contained smooth grain surfaces which appeared to be coated with a glassy phase. The detailed microscopy of these surfaces will be discussed elsewhere in a future publication.

3.2 Bend Strength as a Function of Heat Treatment

From the dilatometric data reported in Figure 2 it is clear that critical temperatures in the shrinkage of the samples are at 200°C, 500°C and 900°C. Consequently, three point bend tests were performed on specimens heated at these temperatures for 1 hour and cooled to room temperature. The results of these tests are presented in Table 1. Ten samples were prepared at each temperature.

It is evident that a large increase in bend strength occurs on heating to 900°C when the glassy surface phases are produced. However, useful increases in strength can be seen to result from a heat treatment at 500°C.

3.3 FTIR Spectroscopy

The relationships between wave number, % transmittance and heat treatment temperature are shown in Figure 3 for the 10A3Y2M2 powder mix. For the powder in the "green" state and after heat treatment and 200°C the FTIR spectra are very similar. Both spectra show deep transmittance valleys at

Table 1. Bend strength (MPA) of the specimens after TST

Temperature of TST, °C	Mix 10A3Y2M2	Mix 10A4Y2M1
200	5·4 ± 0·3	4·6 ± 0·3
500	8·9 ± 0·5	8·9 ± 0·3
900	34·5 ± 3·6	30·9 ± 3·2

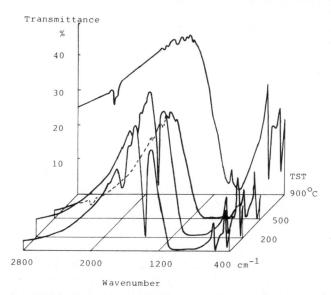

Figure 3. FTIR transmittance spectra of the 10A3Y2M2 powder mix as a function of both wavenumber and TST temperature.

approximately 1385 and 1640 cm⁻¹ with a deep trough (full absorption) in the range 800–1100cm⁻¹. After heat treatment at 500°C and 900°C, the valleys at 1385 and 1640 cm⁻¹ disappear. Two small transmittance valleys occur at approximately 450 and 600 cm⁻¹ for all samples studied. It is interesting to note that these two valleys are both associated with the presence of $SiO_2{}^{3-}$ (495, 600 cm⁻¹) and indicate that all surfaces are oxidised to a detectable amount in the powder mixes. (The valleys are characteristic of both SHS and PCS silicon nitride which have also been studied as pure individual powders). Powder mixes heated to both 500°C and 900°C show a small sharp decrease in the transmittance spectra at approximately 2400 cm⁻¹, with that at 900°C being more pronounced than that at 500°C. These characterise the formation of HCO_3^- (2370 cm⁻¹) and indicate possible reaction with CO_2 in the atmosphere on heating. The valleys at 1385 and 1640 cm⁻¹ have not yet been identified conclusively, but appear to be associated with the hydrated complexes discussed in the previous section.

4. CONCLUSIONS

4.1

It is evident from these preliminary studies that the chemical changes occuring on the powder surfaces on heating, which in turn cause the shrinkages, mass losses and increases in strength, are complicated. The changes involve the loss of physically and chemically combined water, the oxidation of surfaces and possible formation of carbonates and finally the reaction of an active silica-containing surface with very fine active Al_2O_3 to produce an inorganic, glassy binder phase.

4.2

The roles of Y_2O_3 and MgO additions in the complex mixes are not clear but are felt to be very limited because the strengthening phenomenon has been observed without these additions, providing that Al_2O_3 is present.

4.3

These preliminary studies show clearly that it is possible to achieve usable green densities and strengths without the additions of binders. The strengths achieved at 500°C and 900°C are sufficient to permit mechanical handling and machining of compacts.

ACKNOWLEDGMENTS

The authors express their gratitude to the British Council for financial support which allowed the collaboration between Dnepropetrovsk and Manchester to occur. Special thanks are also expressed to Dr. M. A. Hepworth and colleagues at T. and N. Technology Ltd., for discussions and their significant assistance in overpressure sintering of specimens used elsewhere in this work.

REFERENCES

1. OSTRIK, P. N., GASIK, M. M. & POPOV, E. B., in *Adv. Ceramics and P/M Materials,* Proc. Inter. New Bus. and High-Tech. Res. Conf., EAM Institute, Hämeenlinna, Finland, Sept. 23–25, (1990).
2. RAMIS, G., QUINTARD, P., CAUCHETIER, M., BUSCA, G. & LONRENZELLI, V., *J. Amer. Ceram. Soc.,* **73,** 1692, (1989).
3. NAGAI, H., HOKAZONO, S. & KATO, A., *Br. Ceram. Trans. J.,* **90,** 44, (1991).
4. LANGE, F. F., *J. Amer. Ceram. Soc.,* **72,** 3, (1989).
5. BERGSTROM, L. & PUGH, R., *ibid.,* **72,** 103, (1989).
6. KAWAMOTO, M., ISHIZAKI, C. & ISHIZAKI, K., *J. Mater. Sci. Lett.,* **10,** 279, (1991).

Hot Isostatic Pressing of Silicon Nitride and Silicon Carbide using Glass Encapsulation

R. C. PILLER, S. J. FRIEND,* R. W. DAVIDGE

AEA Technology, Harwell Laboratory, Didcot, Oxfordshire, U.K.

J. SLEURS and R. GILISSEN

SCK/VITO, Mol, Belgium

ABSTRACT

Hot Isostatic Pressing using glass encapsulation of green samples has been used to produce dense silicon nitride and silicon carbide. A matrix of HIP conditions was investigated: times — 30, 60 and 120 min; temperatures — 1700, 1800 and 1900°C; pressure — 195 MPa argon. The effect the level of sintering aids (Y_2O_3 and Al_2O_3 for Si_3N_4, C and B for SiC) added to the pure powders has on the properties of the HIPed material was evaluated by measuring the density, hardness and toughness. Fully dense SiC containing no sintering aids was obtained by encapsulation HIPing at 1900°C for 120 min. Si_3N_4 could not be densified without additives present. In this study the addition of SiC whiskers was shown to have little effect on the properties of either material.

1. INTRODUCTION

There are three main techniques of Hot Isostatic Pressing (HIP): Post-HIP of pre-sintered material containing no open porosity; Sinter-HIP, in which a green body is first sintered to closed porosity in the HIP vessel and then given a full HIP treatment; HIP using encapsulation which allows one-step processing of green compacts, the capsules preventing the gas used for pressurisation from entering the pores of the compact. All three processes have areas in which they are best applied [7]. In a previous paper [1] we reported the results from experiments in which three encapsulation techniques were tried: glass tubes, hot-pressed glass compacts and sinter-canning. In this paper we have concentrated on the glass tube encapsulation method [2, 3] but have extended the material and processing parameters. A matrix of HIP conditions has been investigated: times — 30, 60 and 120 min; temperatures — 1700, 1800 and 1900°C; pressure — 195 MPa argon. The effect the level of sintering aids used — Y_2O_3 and Al_2O_3 for Si_3N_4, C and B for SiC — and the incorporation of SiC whiskers has on the properties of the HIPed materials has been evaluated by measuring the density, hardness and toughness of the materials produced across the matrix of conditions.

There are many papers on HIP of Si_3N_4 [4–14], the results from which are as varied as the number of different authors. The main conclusion to be drawn, as Miyamoto *et al.* [8], Takata *et al.* [9] and Tanaka *et al.* [10] all show, is that the properties of the starting powders play a crucial role in achieving full densification, especially when additive free powders are used.

---

*Now BAe, Bristol, U.K.

Pejryd [11], HIPing at temperatures between 1550 and 1750°C, could not obtain dense material unless sintering aids were used. Celis *et al.* [12], investigating the effect of adding ultra fine Si_3N_4 particles to standard powder mixes, found enhanced densification at 1700–1800°C but only if sintering aids were present. Miyamoto *et al.* [8] obtained dense material after encapsulation HIPing pure Si_3N_4 at 1850–1900°C for 1 h. Nezuka *et al.* [13] densified pure Si_3N_4 by HIPing for 3 h at 1900°C and 180 MPa whilst Homma *et al.* [14] went to 2000°C and 200 MPa. Yeheskel *et al.* [15], extrapolating from results from lower temperature work, predict full densification at 1800°C and 250 MPa or at 1600°C and 700 MPa.

Larker *et al.* [16] found a similar dependence on starting powder parameters with SiC. They were able to produce dense material after HIPing at 1850°C, 200 MPa for 1 h without adding sintering aids by refining the particle size of their powders. Likewise, Homma *et al.* [17], working with eight different starting powders and no sintering additives, HIPing at 1900–2000°C and 150 MPa, found that a particle size of 0·6 micron or less was required to produce dense material at 1950°C.

The inclusion of a hard second phase and the effect this has on the densification of the matrix has been discussed by several authors [18–21]. There is a consensus that backstresses developed when the matrix densifies around a non-shrinking inclusion constrains the shrinkage of the matrix and prevents full densification. Zheng *et al.* [22] studied pressureless sintering of Si_3N_4 containing sintering aids, with and without SiC whiskers and platelets. They found that the addition of whiskers inhibited the densification process after treatment at 1700°C for 1 h. Lunberg *et al.* [23] working with Si_3N_4 containing low levels of sintering aids and 25% SiC whiskers obtained similar results and concluded that the composite material required either Hot Pressing or HIPing at 200 MPa in order to densify. Greil *et al.* [24] post-HIPed Si_3N_4 containing SiC particles at 2000°C and 100 MPa and Takemura *et al.* [24] obtained dense pure Si_3N_4 with up to 30% SiC whiskers by HIPing at 1900°C and 180 MPa for 3 h.

The research programme, in which the results presented in this paper form only a part, also included work on the development and use of HIP Maps. For both SiC and Si_3N_4 a consistent set of results over a range of conditions was required in order to have valid input to the model. As pointed out above, this is only possible if a single powder source is used and by doing so we hope to be able to draw meaningful conclusions both from the experimental results and the model predictions.

2. EXPERIMENTAL

The silicon carbide powder used in all experiments was HSC-059 β-grade supplied by Superior Graphite Co., U.S.A. Before use the powder was acid washed, using hydrofluoric acid, in order to reduce the level of SiO_2 present. Carbon was added to the SiC mixtures in the form of glucose, BDH Chemicals Ltd., AnalaR range. Boron was added as amorphous powder, again from BDH Chemicals Ltd.

**Table 1. Properties and impurity levels of
as-received powders**

	Si_3N_4	SiC
Particle size	0·5 μm	0·5 μm
Surface area	12·1 m² g⁻¹	15 + m²g⁻¹
N	—	0·21%
O	0·081%	
C (free)	0·039%	0·85%
Al	0·040%	180 ppm
Fe	0·040%	160 ppm
Cr	—	35 ppm
Ni	—	80 ppm
Ti	—	100 ppm
Mo	—	180 ppm
Mn	—	40 ppm
Zr	—	100 ppm
V	—	150 ppm

The silicon nitride powder was SNE-10 α-grade supplied by UBE Industries Ltd., Japan. The alumina used was AHPS 40 grade supplied by Sumitomo, Japan and the yttria 99·99% pure obtained from Molycorp, U.S.A. The characteristics of the two starting powders are listed in Table 1.

The silicon carbide whiskers were from American Matrix Inc., U.S.A. The diameter of the whiskers ranged from 1 to 3 microns and the length from 20 to 300 microns.

All the powder combinations were prepared using a wet mixing route. The powders were ultrasonically dispersed in distilled water, using Dispex A40 as a dispersing agent. The suspensions were then freeze dried.

Pellets, 8·5 mm in diameter and 10 mm long, of the various powder compositions listed in Tables 2 and 3 were prepared by uniaxial pressing at 78 MPa. These specimens were then isopressed at 175 MPa. Three 'green' specimens of the same composition were encapsulated together in each glass tube. To avoid reactions between the samples and the glass, boron nitride was used as an inert buffer layer. The containers were evacuated at moderate temperatures to ensure complete degassing of the powders, and sealed off. During the project both Vycor (Corning Glass) and fused silica (Heraeus) glass tubes were used. Vycor was initially used for Si_3N_4 with sinter aids, silica for the pure powders and SiC mixes where higher HIP temperatures were anticipated.

For each material one sealed tube (*i.e.* three samples of the same composition) was HIPed at 195 MPa at each combination of time (30, 60 and 120 min) and temperature (1700, 1800 and 1900°C), giving a total of 324 specimens. All runs were done using the National Forge (Europe) Sinter-HIP at SCK/VITO, Belgium.

After HIP, measurements were made of the density of the materials using the water immersion technique. For each combination of composition, HIP

Table 2. List of SiC based compositions and their theoretical densities calculated using a simple 'rule of mixtures' law and the literature values for the densities of the component materials

Mix No.	Theoretical density (g/cm³)	Additives (w/o)		
		B	C	SiC(w)
C1	3·22	—	—	—
C2	3·19	0·18	1	—
C3	3·20	0·36	2	—
C4	3·22	—	—	10
C5	3·19	0·18	1	10
C6	3·20	0·36	2	10

(SiC — 3·22 g cm³
C — 2·00 g cm³
B — 2·37 g cm³).

Table 3(a). Densities of HIPed SiC specimens (All values in g cm⁻³)

Mix No.	Green	1700°C			1800°C			1900°C		
		30'	60'	120'	30'	60'	120'	30'	60'	120'
C1	1·90	2·51	2·60	2·65	2·71	2·78	2·79	3·08	3·11	3·15
C2	1·97	2·59	2·69	2·75	2·82	2·92	2·91	3·14	3·12	3·15
C3	2·02	2·65	2·74	2·84	2·90	2·97	2·97	3·13	3·14	3·13

Table 3(b). Densities of HIPed SiC specimens containing SiC whiskers (All values in g cm⁻³)

Mix No.	Green	1700°C			1800°C			1900°C		
		30'	60'	120'	30'	60'	120'	30'	60'	120'
C4	1·85	2·49	2·58	2·63	2·75	2·79	2·84	3·07	3·11	3·16
C5	1·88	2·56	2·64	2·72	2·75	2·91	2·89	3·11	3·14	3·15
C6	1·97	2·59	2·67	2·77	2·80	2·92	2·95	*	3·14	3·14

temperature and HIP time, three specimens were available; the densities quoted in the tables are averages of the three values obtained.

Optical and Secondary Electron microscopy were used to examine the microstructures. One specimen of each type was sectioned, ground and polished. Examination in the optical microscope gave indications of the amount and distribution of the residual porosity in these samples and also showed how well the whiskers, where present, were distributed throughout the matrix. Etching of the specimens was attempted but no grain size determination using optical microscopy was possible as in all specimens the grains were too small to resolve. Specimens were therefore fractured and examined at higher magnifications in the Scanning Electron Microscope. From observations of the fracture surfaces some appreciation of structure — grain size, grain growth, porosity and whisker distribution — could be obtained, but no quantitative determination of grain size was attempted.

Vickers Hardness values were measured on the polished sections. A Vickers-Armstrong machine was used for all tests, with an indentation load of 20 kg. One specimen from each combination was indented five times and the H_v values quoted in the tables for each combination were averages of these directly measured values.

Fracture Toughness values were determined using the Indentation Fracture technique in which the lengths of cracks emanating from the corners of a Vickers indentation are measured. There are several different equations which can be used to calculate K_{1c} values from these crack lengths, that of Marshall and Evans [26] has been used in this report.

$$K_{1C} = 0.036 \ E^{0.4} \ P^{0.6} \ a^{-0.7} \ (c/a)^{-1.5}$$

where;

E — Young's modulus (310 GPa used for all Si_3N_4 materials)
    (460 GPa used for all SiC materials)
P — indentation load (in kg)
a — half length of diagonal (in mm)
c — crack length from centre of indent (in mm).

3. RESULTS AND DISCUSSION

3.1 Silicon Carbide

The theoretical densities for all the SiC mixes are given in Table 2. The measured densities for mixes in which the level of additives was varied are given in Table 3(a) and for the corresponding mixes containing whiskers in Table 3(b). The variation in density after HIP was found to be much greater than that of the Si_3N_4 specimens. This is to be expected given the higher pressureless sintering temperature for SiC. The bar chart, Figure 1, shows graphically the effect that the level of additives used has on the percentage of the theoretical density achieved over the full range of conditions. Densification to near theoretical values is only achieved after HIP at 1900°C. Increased time at temperature produces only a small increase in density. The addition of C and B increases the amount of densification occuring at all temperatures, but only by a few percent. At 1900°C high densities are achieved for all compositions including the pure SiC powder.

In Figure 2 a bar chart of the effect on density of adding whiskers to the mixes is shown. In general the addition of whiskers appears to have a minimal effect on the density achieved. With pure SiC the variations seen in the densities of specimens with and without whiskers are neither consistently plus or minus nor significantly larger than the error in the density measurements themselves. For the mix containing the full level of additives, the presence of whiskers slightly retards the densification process, especially in the specimens which have not attained full density. In the dense specimens, *i.e.* those HIPed

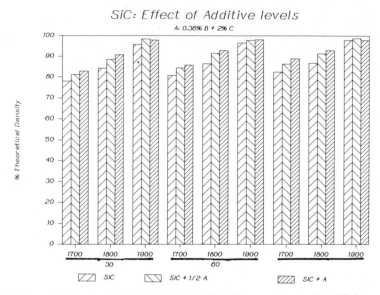

Figure 1. Bar graph showing the effect of the level of additives on the density of SiC over the full range of HIP temperatures and times.

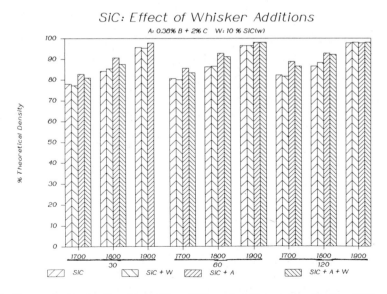

Figure 2. Bar graph showing the effect of the addition of whiskers on the density of SiC over the full range of HIP temperatures and times.

(a)   (b)   (c)

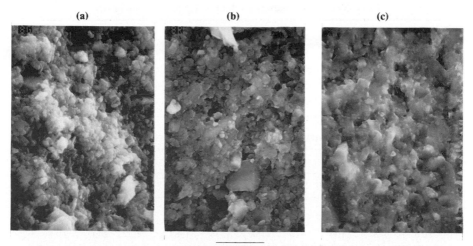

5 microns

**Figure 3. Scanning electron micrographs of fracture surfaces in SiC + 2% C + 0·36% B +
10% SiC(w),
(a) HIPed 195 MPa, 1700°C, 120 min.
(b) HIPed 195 MPa, 1800°C, 120 min.
(c) HIPed 195 MPa, 1900°C, 120 min.**

(a)   (b)

15 microns

**Figure 4. Scanning electron micrographs of fracture faces of specimens HIPed at 195 MPa for
120 min at 1900°C.
(a) SiC + 1% C + 0·18% B.
(b) SiC + 1% C + 0·18% B + 10% SiC(w).**

at 1900°C for either 60 or 120 min, the addition of whiskers has no effect on
the density achieved.

Examination of fracture surfaces in specimens from each of the six mixes
HIPed at 1900°C show a coarsening of the microstructure as the level of
additives increases. This is also apparent in Figure 3 in which the
microstructures of the SiC with full additive level and 10% whiskers after HIP

Table 4(a). Vickers hardness values of HIPed SiC specimens

Mix No.	1700°C			1800°C			1900°C		
	30'	60'	120'	30'	60'	120'	30'	60'	120'
C1	721	844	975	1193	1245	1201	2397	2259	2412
C2	784	965	990	1150	1479	1606	2373	2150	2405
C3	790	1059	1213	1492	1710	1767	2008	2442	2329

Table 4(b). Vickers hardness values of HIPed SiC specimens containing SiC whiskers

Mix No.	1700°C			1800°C			1900°C		
	30'	60'	120'	30'	60'	120'	30'	60'	120'
C4	629	762	849	1077	1145	1450	2193	2269	2375
C5	720	873	1028	1221	1402	1299	2228	2292	2428
C6	625	1104	910	1154	1234	1583	*	2497	2368

Table 5(a). Fracture toughness values of HIPed SiC specimens

Mix No.	1700°C			1800°C			1900°C		
	30'	60'	120'	30'	60'	120'	30'	60'	120'
C1	4·10	2·75	3·65	3·45	3·18	3·31	3·29	2·41	2·40
C2	—	3·77	3·79	4·59	3·27	3·60	2·59	3·31	2·46
C3	4·44	3·34	3·29	4·52	3·15	3·11	2·72	3·56	3·75

Table 5(b). Fracture toughness values of HIPed SiC specimens containing SiC whiskers

Mix No.	1700°C			1800°C			1900°C		
	30'	60'	120'	30'	60'	120'	30'	60'	120'
C4	—	—	—	—	3·28	3·50	2·86	3·13	3·12
C5	—	—	3·06	4·85	4·15	3·87	3·90	3·54	2·92
C6	—	4·26	5·46	6·28	4·30	4·77	*	3·32	3·23

at the three different temperatures for maximum time are compared. However, the extent of this grain growth is no more than a factor of × 3. There is no evidence of the exaggerated grain growth found in pressureless sintered materials of the same composition. Comparison of micrographs taken of fracture surfaces in the SiC with half additives and the same mix with 10% whiskers, Figures 4(a) and 4(b), show that the whiskers survive after processing at the highest temperature (1900°C) and longest time (120 min). Other than the presence of the whiskers, the grain structure in these two specimens is very similar.

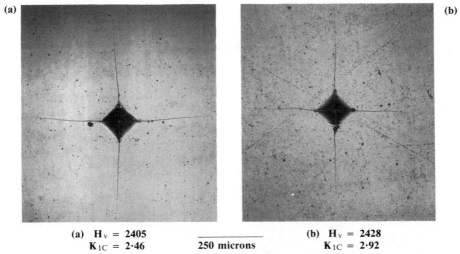

(a) H_v = 2405
      K_{1C} = 2·46   250 microns   (b) H_v = 2428
                                                                       K_{1C} = 2·92

Figure 5. Optical micrographs of Vickers hardness indents and the cracks emanating from the indent corners from which fracture toughness values were determined.
(a) SiC + 1% C + 0·18% B.
(b) SiC + 1% C + 0·18% B + 10% SiC(w).

For all the SiC compositions the Vickers Hardness values, given in Tables 4(a) and 4(b), reflect the range of densities obtained over the matrix of conditions. The denser the material the harder it is. The best results obtained, 2200–2400 H_v in the materials HIPed at 1900°C, are consistent with best literature values. There is no significant difference due to the level of additives present, neither can it be said that the addition of whiskers induces any change in H_v. Micrographs of typical indentations are shown in Figure 5.

The Fracture Toughness (K_{1c}) values, given in Tables 5(a) and 5(b), require careful interpretation. In the low density materials the apparent toughening is due to the high porosity. In the dense materials the values obtained are low but the addition of additives generally causes a slight increase. The low values of toughness measured in these specimens is most likely a consequence of the equiaxed grain structure and reflects the absence of the large acicular grains commonly seen in pressureless sintered materials. This premise is reinforced by the observation that the addition of the acicular whiskers does give a small increase in toughness. The slight increase seen when additives are present is most likely due to the small increase in grain size found in these materials over that found in the pure SiC.

3.2 Silicon Nitride

The theoretical densities, calculated using a simple rule of mixtures, of the Si_3N_4 compositions used in this series of experiments are given in Table 6. The densities obtained over the whole matrix of compositions and conditions are given in Table 7(a) and in 7(b) for those compositions incorporating whiskers.

Table 6. List of Si_3N_3 based compositions and their
theoretical densities calculated using a simple 'rule of
mixtures' law and the literature values for the densities
of the component materials

Mix No.	Theoretical density (g/cm³)	Additives (w/o)		
		Al_2O_3	Y_2O_3	SiC(w)
N1	3·18	—	—	—
N2	3·21	0·5	1·25	—
N3	3·23	1	2·5	—
N4	3·29	2	5	—
N5	3·18	—	—	10
N6	3·23	2	5	10

$(Si_3N_4 — 3·18$ g cm³
$SiC — 3·22$ g cm³
$Al_2O_3 — 3·98$ g cm³
$Y_2O_3 — 5·01$ g cm³)

Table 7(a). Densities of HIPed Si_3N_4 specimens

Mix No.	Green	1700°C			1800°C			1900°C		
		30'	60'	120'	30'	60'	120'	30'	60'	120'
N1	1·66	2·06	2·04	2·06	2·19	2·54	*	2·59	2·72	2·77
N2	1·63	3·18	3·22	3·20	3·20	3·20	3·21	3·20	3·20	3·20
N3	1·65	3·20	3·23	3·23	3·22	3·22	3·22	3·23	3·23	3·23
N4	1·68	3·22	3·26	3·26	*	3·25	3·22	3·24	3·25	3·24

Table 7(b). Densities of HIPed Si_3N_4 specimens containing SiC whiskers

Mix No.	Green	1700°C			1800°C			1900°C		
		30'	60'	120'	30'	60'	120'	30'	60'	120'
N5	1·61	2·18	2·18	2·23	2·33	2·46	2·62	2·83	2·90	3·00
N6	1·64	3·20	3·23	3·24	3·23	3·23	3·24	3·23	3·23	3·23

The bar chart in Figure 6, shows the effect changing the level of additives —
full, half and quarter levels — has on the percentage of the theoretical density
achieved. For all the conditions used the presence of additives is necessary to
achieve densification. The amount of additive used, however, has little effect
on the density achieved. A quarter of the 'standard' level is sufficient to obtain
almost fully dense material.

The bar chart in Figure 7 shows the effect on density of adding 10% SiC
whiskers to the pure Si_3N_4 and to the mix containing the full level of additives.
In the pure powder, for nearly all combinations of time and temperature, the

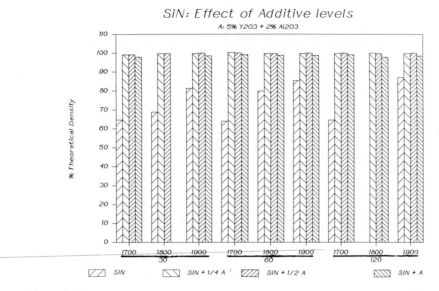

Figure 6. Bar graph showing the effect of the level of additives on the density of Si₃N₄ over the full range of HIP temperatures and times.

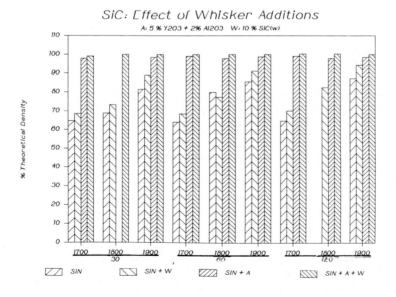

Figure 7. Bar graph showing the effect of the addition of whiskers on the density of Si₃N₄ over the full range of HIP temperatures and times.

(a)   (b)   (c)

5 microns

Figure 8. Scanning electron micrographs of fracture surfaces in Si$_3$N$_4$ + 2% Al$_2$O$_3$ + 5% Y$_2$O$_3$ + 10% SiC(w).
(a) HIPed 195 MPa, 1700°C, 120 min.
(b) HIPed 195 MPa, 1800°C, 120 min.
(c) HIPed 195 MPa, 1900°C, 120 min.

whiskers promote densification. This is most likely due to the addition, along with the whiskers, of extra impurities such as SiO$_2$ which increase the amount of liquid phase created. In the mix containing the full level of additives, where increasing the amount of glass phase present would be superfluous, the addition of whiskers causes a much smaller increase in the densities achieved and this increase in only evident if the theoretical densities are compared; the measured densities actually show a decrease on addition of whiskers.

From these density results, the minimum HIPing conditions necessary to produce dense Si$_3$N$_4$ from powder containing additives, with and without whiskers, would be any time between 0·5 and 1 h at 1700°C at a pressure of 195 MPa. All times at higher temperatures produced material of similar densities. Optical micrographs of polished sections of the specimens show that although of similar density the amount and distribution of the porosity in these specimens can vary. Of all the high density specimens, those containing both full additives and whiskers appear the most porous, this could be due to geometrical restrictions on densification in the vicinity of whiskers.

The scanning electron micrographs in Figure 8 indicate that the grain structure in the dense materials is very similar, the main effect of higher additive levels is to increase the amount and size of the β grains. This increase in grain size is no larger than a factor of three. The micrographs all show a mixture of equiaxed and acicular grains. The starting Si$_3$N$_4$ powder is α-phase and the acicular grains are the β-phase.

For all compositions containing whiskers, their distribution throughout the specimens has been found to be very homogeneous. The whiskers survive even

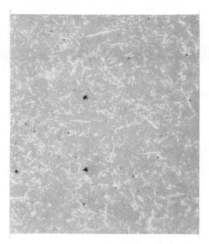

Figure 9. Optical micrograph of Si₃N₄ + 2% Al₂O₃ + 5% Y₂O₃ + 10% SiC(w) HIPed at 195 MPa for 120 min at 1900°C.

Table 8(a). Vickers hardness values of HIPed Si₃N₄ specimens

Mix No.	1700°C			1800°C			1900°C		
	30'	60'	120'	30'	60'	120'	30'	60'	120'
N1	324	251	—	420	672	*	818	876	949
N2	1893	1770	1754	1786	—	1734	1773	1815	1752
N3	1992	1699	1663	1802	—	1578	1687	1688	1595
N4	1788	1645	1659	*	1579	1559	1575	1574	1617

Table 8(b). Vickers hardness values of HIPed Si₃N₄ specimens containing SiC whiskers

Mix No.	1700°C			1800°C			1900°C		
	30'	60'	120'	30'	60'	120'	30'	60'	120'
N5	329	367	445	338	669	843	1031	1203	1323
N6	1810	1743	1699	1715	1624	1586	1719	1672	1649

after HIPing at the highest temperature, 1900°C, for the longest time, 120 min, and in the presence of the highest amount of liquid phase, as shown in Figure 9.

The Vickers hardness values, H_v, given in Tables 8(a) and 8(b), reflect the consistently high densities obtained in the compositions with additives and the low densities of the two mixes without. The general trend appears to indicate that the material which has been HIPed at the lowest temperature, 1700°C, for the shortest time, 30 min, is the hardest. The implication is that the slight coarsening of the grain structure observed in the SEM examination of the

(a)   **(b)**

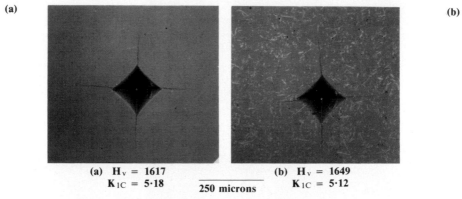

(a) $H_V = 1617$ (b) $H_V = 1649$
     $K_{1C} = 5{\cdot}18$ ───── 250 microns $K_{1C} = 5{\cdot}12$

Figure 10. Optical micrographs of Vickers hardness indents and the cracks emanating from the
indent corners from which fracture toughness values were determined.
(a) $Si_3N_4 + 2\%\ Al_2O_3 + 5\%\ Y_2O_3$.
(b) $Si_3N_4 + 2\%\ Al_2O_3 + 5\%\ Y_2O_3 + 10\%\ SiC(w)$.

Table 9(a). Fracture toughness values of HIPed Si_3N_4 specimens

Mix No.	1700°C			1800°C			1900°C		
	30'	60'	120'	30'	60'	120'	30'	60'	120'
N1	—	—	—	—	3·09	*	3·63	2·76	2·84
N2	3·56	3·17	4·14	4·39	—	4·32	3·43	3·45	3·73
N3	3·91	4·39	4·97	5·40	—	4·63	4·15	4·40	4·44
N4	4·50	4·60	5·72	*	5·12	5·33	5·34	5·36	5·18

**Table 9(b). Fracture toughness values of HIPed SiN_4 specimens containing SiC
whiskers**

Mix No.	1700°C			1800°C			1900°C		
	30'	60'	120'	30'	60'	120'	30'	60'	120'
N5	—	—	—	—	—	3·42	2·17	2·88	2·64
N6	5·34	5·14	5·07	5·03	5·63	5·57	5·22	4·88	5·12

fracture surfaces is responsible for a reduction in the hardness of these
materials. This explanation is also consistent with the observation that there is
a slight increase in hardness, seen over the whole range of conditions, as the
level of additives, and hence the amount of coarsening, is decreased. The
presence of whiskers slightly hardens the material. Micrographs of typical
indentations are shown in Figure 10.

The mixes, without additives, HIPed at the lower temperatures produced
material which was too porous to give meaningful toughness measurements
using the indentation technique. From the results given in Tables 9(a) and 9(b)
the main observation to be made is that the toughness decreases with

decreasing level of additives. As the level of additives decreases so does the amount of liquid phase created. Consequently the amount and size of β-Si_3N_4 grains, precipitated from the liquid phase after the dissolution of the α-phase, is reduced. The β-grains are acicular and the higher the percentage present, *i.e.* the greater the $\alpha \rightarrow \beta$ conversion, the tougher the material. The results also show the addition of whiskers produces neither significant nor consistent toughening of the matrix material. This is to be expected considering the similarity in morphology of the whiskers and the β-Si_3N_4 grains.

4. CONCLUSIONS

4.1 SiC

— For all compositions studied, near theoretical density is achieved when HIPing at 1900°C. At the lower temperatures, 1700 and 1800°C, densification is incomplete.
— At 1900°C, where conditions are right for densification, the addition of C and B as sinter aids is not necessary. At the lower temperatures denser materials are obtained when additions of C and B had been made.
— Increased time at temperatures gives higher densities. The presence of sintering aids appears to increase the rate of densification.
— For the dense materials the grain sizes observed, over the range of compositions and times at 1900°C, varies by no more than a factor of × 3.
— The hardness and toughness values measured on the dense SiC materials are comparable with the best literature values produced using the same test methods.
— The use of sintering aids has no significant effect on the hardness or toughness values of dense material.
— The addition of SiC whiskers slightly retards the densification.
— The presence of whiskers in dense materials has no significant effect on the hardness or toughness values obtained.

4.2 Si_2N_4

— Pure Si_3N_4 powder does not densify under any of the conditions attempted in this series of experiments.
— The use of sintering aids is necessary to achieve complete densification. When these additives are present, near theoretically dense material is produced by HIPing even at the lowest temperature, 1700°C, and the shortest time, 30 min.
— The actual level of additives required to achieve complete densification has not been established but the lowest level used in this series of experiments, 1·25% Y_2O_3 + 0·5% Al_2O_3, is sufficient. It may be possible to use even lower levels.
— The grain sizes observed, over the whole range of compositions, temperatures and times, varies by no more than a factor of × 3.
— The higher the amount of additives used the lower the hardness and higher the toughness values measured on the HIPed material.

— The microstructure, and consequently the properties, of the HIPed material can alter significantly according to the amount and viscosity of the liquid phase present at high temperatures. The amount of liquid phase produced is directly proportional to the level of sintering aids in the starting powder. The lower the amount of liquid phase present the slower and more incomplete the $\alpha \rightarrow \beta$ transition and it is the ratio of α/β phase present in the final material which determines the properties. Generally, the lower the α/β phase ratio the lower the hardness but higher the toughness. Increasing the HIP temperature and time also lowers the α/β phase ratio.

— The addition of SiC whiskers has no significant effect on the hardness or toughness values obtained.

4.3 Whiskers

This set of experiments gave results in which the addition of SiC whiskers to either SiC or Si_3N_4 powders gave final products with no improvement in properties. The whiskers were well distributed in all materials and survived the harshest HIP conditions. They appear to be well bonded within the matrix, the microscopy on the fracture surfaces shows no evidence for whisker pull-out and the cracks generated in the hardness tests are not deflected by the presence of whiskers, they often pass straight through them. The interfacial bonding of the whiskers in these materials is therefore too good. Weak interfaces are required to obtain the most effect from the whiskers and it is possible that if interfacial engineering — coating the whiskers with a material which would inhibit bonding — were attempted, improvements in the properties would result.

ACKNOWLEDGMENTS

Financial support for this work came from the CEC, within the EURAM Advanced Materials Programme under Contract No. MA1E/002/C. Grateful thanks are also due to our Commercial partners in this project; National Forge (Europe), BP International Ltd., Cerametal (Luxembourg), ICI plc. Advanced Materials and RTZ Chemicals Ltd.

REFERENCES

1. PILLER, R. C., FRIEND, S. J. & DENTON, I. E., *Brit. Ceram. Proc.,* **45**, 33, (1990).
2. LARKER, H. T., ADLERBORN, J. & BOHMAN, H., SAE Tech. Paper No. 770335, (1977).
3. LARKER, H. T., *Mat. Sci. Res.,* **17**, 571, (1984).
4. WILLS, R. R., BROCKWAY, M. C., McCOY, L. G. & NEISZ, D. E., *Ceram. Eng. and Sci. Proc.,* **1B**, 534, (1980).
5. PEJRYD, L., *Proc. Int. Conf.,* *"Hot Isostatic Pressing — Theories and Applications,"* Lulea, Sweden, (1987), 307.
6. INGESTRÖM, N. & EKSTRÖM, K., *Proc. Int. Conf.,* *"Hot Isostatic Pressing — Theories and Applications,"* Lulea, Sweden, June, (1987), 307.
7. DAVIDGE, R. W., SLEURS, J. & BUEKENHOUT, L., in *Proc. Int. Conf. on Hot Isostatic Pressing of Materials: Applications and Developments,* Antwerp, Belgium, April, (1988).

8. MIYAMOTO, Y., TANAKA, K., SHIMADA, M. & KOIZUMI, M., *Proc. 2nd Int. Symp. Ceramic Materials and Components for Engines,* Lubeck-Travemunde, Germany, April, (1986).
9. TAKATA, H., MARTIN, C. & ISHIZAKI, K., *J. Ceram. Soc. Japan, Int. Ed.,* **96**, 874, (1988).
10. TANAKA, I., PEZZOTTI, G., OKAMOTO, T. & MIYAMOTO, Y., *Ceram. Eng. Sci. Proc.,* **10**, 817, (1989).
11. PEJRYD, L., *Advanced Ceram. Mat.,* **3**, 403, (1981).
12. CELIS, P., ISHIZAKI, K., WATARI, K., MIYAMOTO, A., FUYUKI, T. & YAMADA, Y., *Proc. Int. Conf. on Hot Isostatic Pressing of Materials: Applications and Developments,* Antwerp, Belgium, April, (1988).
13. NEZUKA, K., MIYAMOTO, Y. & KOIZUMI, K., *Proc. Int. Conf., "Hot Isostatic Pressing — Theories and Applications,"* Lulea, Sweden, June, (1987), 359.
14. HOMMA, K., OKADA, H., FUJIKAWA, T. & TATUNO, T., *J. Ceram. Soc. Japan, Int. Ed.,* **95**, 195, (1987).
15. YEHESKEL, O., GEFEN, Y. & TALIANKER, M., *Mat. Sci. and Eng.,* **78**, 209, (1986).
16. LARKER, H. T., HERMANSSON, L. & ADLERBORN, J., *Ind. Ceram.,* **8**, 17, (1988).
17. HOMMA, K., YAMAMOTO, F. & OKADO, H., *J. Ceram. Soc. Japan,* **95**, 195, (1987).
18. ASHBY, M. F., BAHK, S., BERK, J. & TURNBALL, D., *Prog. in Mat. Sci.,* **25**, 1, (1980).
19. RAJ, R. & BORDIA, R. K., *Acta Met.,* **32**, 1003, (1984).
20. HSUEH, C. H., EVANS, A. G., CANNON, R. G. & BROOK, R. J., *Acta Met.,* **34**, 327, (1986).
21. CLEGG, W., ALFORD, McN. N. & BIRCHALL, J. D., *Brit. Ceram. Proc., Engineering with Ceramics,* **2**, London, 247, (1986).
22. ZHENG, X. Y., ZENG, F. F., POMEROY, M. J. & HAMPSHIRE, S., *Brit. Ceram. Proc.,* **45**, 187, (1990).
23. LUNDBERG, R., NYBERG, B., WILLIANDER, K., PERSSON, M. & CARLSSON, R., *Proc. Int. Conf., "Hot Isostatic Pressing — Theories and Applications,"* Lulea, Sweden, June, (1987).
24. GREIL, P., PETZOW, G. & TANAKA, H., *Ceram. International,* **13**, 19, (1987).
25. TAKEMURA, H., MIYAMOTO, Y. & KOIZUMI, K., *Proc. Int. Conf., "Hot Isostatic Pressing — Theories and Applications,"* Lulea, Sweden, June, (1987).
26. MARSHALL, D. B. & EVANS, A. G., *J. Amer. Ceram. Soc.,* **64**, C182, (1981).

Author Index

Subject Index